Model Systems Engineering Documents for Adaptive Signal Control Technology (ASCT) Systems

August 2012

U.S. Department of Transportation
Federal Highway Administration

FHWA-HOP-11-027

NOTICE

This document is disseminated under the sponsorship of the U.S. Department of Transportation in the interest of information exchange. The U.S. Government assumes no liability for the use of the information contained in this document. This report does not constitute a standard, specification, or regulation.

The U.S. Government does not endorse products or manufacturers. Trademarks or manufacturers' names appear in this report only because they are considered essential to the objective of the document.

QUALITY ASSURANCE STATEMENT

The Federal Highway Administration (FHWA) provides high-quality information to serve Government, industry, and the public in a manner that promotes public understanding. Standards and policies are used to ensure and maximize the quality, objectivity, utility, and integrity of its information. FHWA periodically reviews quality issues and adjusts its programs and processes to ensure continuous quality improvement.

Technical Report Documentation Page

1. Report No. FHWA-HOP-11-027	2. Government Accession No.	3. Recipient's Catalog No.	
4. Title and Subtitle Model Systems Engineering Documents for Adaptive Signal Control Technology Systems - Guidance Document		5. Report Date August 2012	
		6. Performing Organization Code	
7. Authors DKS Associates: Kevin Fehon, Mike Krueger, Jim Peters Federal Highway Administration: Richard Denney, Paul Olson, Eddie Curtis		8. Performing Organization Report No.	
9. Performing Organization Name and Address Science Applications International Corporation 1710 SAIC Drive, MS T1-12-3, McLean, Virginia 22102 DKS Associates, 1970 Broadway, Suite 740, Oakland CA 94612		10. Work Unit No. (TRAIS)	
		11. Contract or Grant No. DTFH61-06-D-00005	
12. Sponsoring Agency Name and Address Federal Highway Administration – HOTM 1200 New Jersey Avenue, SE. Washington, DC 20590		13. Type of Report and Period Covered Project Report, December 2010 – August 2012	
		14. Sponsoring Agency Code FHWA - HOP	
15. Supplementary Notes Eddie Curtis, FHWA Office of Operations– Atlanta Rick Denney, Paul Olson, FHWA National Resource Center			
16. Abstract Model Systems Engineering Documents for Adaptive Signal Control Technology Systems is intended to provide guidance for professionals involved in developing systems engineering documents covering the evaluation, selection and implementation of adaptive signal control technology systems.			
17. Key Words Systems engineering, adaptive traffic signals, ASCT, concept of operations, system requirements, verification, validation, intelligent transportation system		18. Distribution Statement No restrictions.	
19. Security Classif. (of this report) Unclassified	20. Security Classif. (of this page) Unclassified	21. No of Pages 276	22. Price N/A

Form DOT F 1700.7 (8-72)   Reproduction of completed page authorized.

Acknowledgements

The following people (from State, regional and local transportation agencies) generously gave their time to review drafts of this document and provided valuable comments that helped to greatly improve its quality.

Glenn Hansen, City of Omaha, NE

Mike Whiteaker, City of Bellevue, WA

James Sturdevant, Indiana DOT

Val Rader, Alaska DOT

Leo Almanzer, Meadowlands Commission, NJ

Eric Graves, City of Alpharetta, GA

Ray Web, Mid-America Regional Council, (MARC)

Kevin Lacy, North Carolina DOT

Mark Taylor, Utah DOT

Dongho Chang, City of Seattle, WA

Matt Meyers, Missouri DOT

Karen Aspelin, Consultant representing Bernalillo County, NM

TABLE OF CONTENTS

EXECUTIVE SUMMARY 1

A. PREAMBLE 3

    Why Use Systems Engineering? 3

    Purpose of This Document 3

    Why Consider ASCT? 4

    How to use this document 5

B. OVERVIEW OF THE PROCESS 8

    Assembling your documents 10

C. THE SYSTEMS ENGINEERING DOCUMENTS 11

    Overview 11

    Responsibilities 11

    Concept of Operation 12

    Requirements 13

    Verification Plan 14

    Validation Plan 14

    Procurement Plan 15

        Request for Information (RFI) 15

        Industry Review of Requirements 16

        Request for Qualification (RFQ) 16

        Request for Proposal (RFP) 16

        Low-Bid 17

        Verification Plan 18

        Market Research Approach 18

        Procurement and Verification references 19

        Sole Source Procurement 19

D. CONCEPT OF OPERATIONS GUIDANCE 21

    Logical Workflow in Preparing the Concept of Operations 21

Chapter 1: Scope	22
1.1 Document Purpose and Scope	22
1.2 Project Purpose and Scope	22
1.3 Procurement	22
Chapter 2: Referenced Documents	22
Chapter 3: User-Oriented Operational Description	22
3.1 How Does the Existing System Work?	23
Network Characteristics	23
Traffic Characteristics	23
Signal Grouping	23
Land Use	23
Operating Agencies	23
Existing Architecture and Infrastructure	24
3.2 What are the Limitations of the Existing System?	24
3.3 How Should the System be Improved?	24
3.4 Statement of Objectives for the Improved System	24
Smooth Flow	25
Maximize Throughput	25
Access Equity	25
Manage Queues	25
Variable Objectives	26
Maximize Isolated Intersection Efficiency	26
3.5 Description of Strategies to be Applied by the Improved System	26
Provide A Pipeline	26
Distribute Phase Splits	27
Manage Queues	27
Variable Strategies	27
3.6 Alternative Non-Adaptive Strategies Considered	27
Traffic Responsive	28

   Complex Coordination Features 28

 Chapter 4: Operational Needs 30

  4.1 *Adaptive Strategies* 30

   Traditional Traffic Operational Strategies 31

   Adaptive Coordination and Control Techniques 31

  4.2 *Network Characteristics* 33

   Number of Signals 33

   Size of Groups 33

   Flexible Groups 33

   Separation of Groups 36

   Relationship of Groups to Other Groups or Facilities 36

   Jurisdictional Relationships 36

  4.3 *Institutional and System Boundaries* 36

   Crossing Arterial Coordination 37

   Adaptive Coordination Strategies 38

  4.4 *Security* 38

  4.5 *Queuing Interactions* 38

  4.6 *Pedestrians* 39

  4.7 *Non-Adaptive Situations* 40

   Adaptively and Automatically 40

   Detect Conditions 40

   Schedule Operation 40

   Operator Override 41

  4.8 *System Responsiveness* 41

   Small Shifts in Demand 41

   Large Shifts in Demand 41

   Response Time 41

  4.9 *Complex Coordination and Controller Features* 42

   Multiple Phase Service 42

       Early Release of Hold 44

       Hold the Position of Uncoordinated Phrases 44

       Late Phase Introduction 44

4.10 Monitoring and Control 44

4.11 Performance Reporting 44

       Performance Measurement 45

       System Operation 45

       Mobility Objectives 45

       Real-Time Logging 45

       Data Storage 46

       External Interfaces 46

       Use Historical Data to Recreate Events 46

4.12 Failure Notification 46

       Report Directly 46

       Report to Interfaced System 46

4.13 Preemption and Priority 47

       Railroad Preemption 47

       Emergency Vehicle Preemption 47

       Transit Priority 47

       Preemption / Priority Frequency 47

4.14 Failure and Fallback Modes 48

4.15 Definition and Application of Constraints 48

       Infrastructure 48

       Management and human Resources 48

       Financial Constraints 49

       Complexity 49

       People 49

       Hardware and Software Constraints 50

       Schedule 50

        5   Chapter 5. Envisioned Adaptive System Overview 50

        6   Chapter 6. Adaptive Operational Environment 53

            6.1   Operational Environment 53

            6.2   Physical Environment 53

      Chapter 7: Adaptive Support Environment.. 53

      Chapter 8: Proposed Operational Scenarios Using an Adaptive System........ 54

            8.1   How to Construct a Scenario 54

E. SYSTEM REQUIREMENTS GUIDANCE 56

    1   Scope of System or Sub-system (Chapter 1 of the System Requirements).................. 57

    2   References (Chapter 2 of the System Requirements)... 57

    3   Requirements (Chapter 3 of the System Requirements) 57

    4   Verification Methods (Chapter 4 of the System Requirements)................................ 58

    5   Supporting Documentation (Chapter 5 of the System Requirements) 59

    6   Traceability Matrix (Chapter 6 of the System Requirements) 59

F. VERIFICATION PLAN GUIDANCE 60

    1   Purpose of Document (Chapter 1 of the Verification Plan)....................................... 61

    2   Scope of Project (Chapter 2 of the Verification Plan) 61

    3   Referenced Documents (Chapter 3 of the Verification Plan) 61

    4   Conducting Verification (Chapter 4 of the Verification Plan) 61

    5   Verification Identification (Chapter 5 of the Verification Plan) 62

G. VALIDATION PLAN GUIDANCE 63

    1   Purpose of Document (Chapter 1 of the Validation Plan) 64

    2   Scope of Project (Chapter 2 of the Validation Plan).. 64

    3   Referenced Documents (Chapter 3 of the Validation Plan)....................................... 64

    4   Conducting Validation (Chapter 4 of the Validation Plan) 64

    5   Validation Identification (Chapter 5 of the Validation Plan) 65

APPENDICES

Appendix A Document Templates

Appendix B Concept of Operations Table of Sample Statements

Appendix C Concept of Operations Example Scenarios

Appendix D System Requirements Table of Sample Requirements

LIST OF FIGURES

Figure 1. Sections of this Document 6

Figure 2. Should You Consider Adaptive Control? 8

Figure 3. Overview of Systems Engineering Flow Chart for Asct System Definition 9

Figure 4. Relationships Between this Guide and Final Documents 12

Figure 5. Best Value Supported by SE Analysis 17

Figure 6. Low-Bid Supported by SE Analysis 18

Figure 7. Market Research / Low-Bid Approach 19

Figure 8. Example of Flexible Grouping 35

Figure 9. Sample System Block Diagram 52

LIST OF TABLES

Table 1. Categorization of Requirements 13

Table 2. Example Traceability Matrix 59

Glossary

ASC	Adaptive Split Control
ASCT	Adaptive Signal Control Technology
Child	Child requirement that is associated with an over-arching requirement called a parent requirement.
CIC	Critical Intersection Control
CFR	Code of Federal Regulations
COTS	Commercially available Off-The-Shelf. This is an FAR term defining a non-developmental item of supply that is both commercial and sold in substantial quantities in the commercial marketplace, and that can be procured or utilized under government contract in precisely the same form as available to the general public.
FAR	Federal Acquisition Regulation
ICM	Integrated Corridor Mobility
ITS	Intelligent Transportation System
Low-bid	Contract awarded to the "lowest responsible bidder". Bid is based on a complete set of plans and specifications that precisely defines the facilities to be built.
Natural cycle length	The cycle length at which an intersection would run with minimum overall delay.
Parent	Parent requirement that has associated sub-requirements called child requirements.
PIF	Public interest finding
Real-time	Activity that occurs simultaneously with or very soon after an event. For example, real-time control involves taking action based on measurements immediately after the measurement is completed.
Resonant cycle length	A cycle length that accommodates good two-way progression.
RFI	Request for information
RFP	Request for proposal
RFQ	Request for qualifications
Synch point	Reference point in signal phase or cycle used to synchronize operation of adjacent signals.
TBC	Time base coordination
TMC	Transportation Management Center
TOC	Traffic Operations Center
TOD	Time of Day
TRPS	Traffic Responsive Pattern Selection

Executive Summary

The vision of the Every Day Counts Adaptive Signal Control Technology (ASCT) Initiative is to mainstream the use of adaptive signal control technology. "Mainstream the use of" suggests that when traffic conditions, agency needs, resources and capability support the use of ASCT it should be implemented. These model systems engineering documents support the process of exploring the need for ASCT and articulating a set of requirements that enable agencies to specify, select, implement and test adaptive signal control technology. Over the last two decades a significant number of adaptive systems have been deactivated well before the end of their useful life due either to a lack of adequate resources or agency capability to support system operation and maintenance, or in some cases a failure to properly align agency and system operations objectives. The risks associated with ASCT implementation are significant.

These documents will assist agencies to apply the systems engineering process in a manner that is commensurate with the scale of the project, in order to substantially reduce the level of effort and address many of the risks associated with procurement of ASCT. The process will also help an agency confirm that its expectations are realistic and achievable before committing to a system. Federal regulation (23 CFR 940.11) requires that any ITS project funded in whole or in part with Highway Trust Funds must be based on a systems engineering analysis, and that the level of effort should be commensurate with the scale of the project. By following the guidance provided in this document and using the model documents, an agency will be able to comply with this regulation while at the same time improving the likelihood of a successful ASCT implementation.

These model documents will provide valuable support to agencies at two distinctly different stages of project development.

- During the planning phase, the preparation of a Concept of Operations will clarify the agency's and other stakeholders' needs for an ASCT system that supports the mobility, air quality and other transportation objectives of the region. This will provide sufficient information to clearly define a project with an appropriate schedule and budget.
- During the project phase, the Concept of Operations can be used to develop system requirements to a level of detail sufficient to successfully procure an ASCT system, and then confirm the system successfully meets the agency's overall objectives.

This document helps accomplish the tasks of clarifying objectives, identifying needs and defining constraints by leading the reader through a series of questions. The outcome of selecting and tailoring the sample responses will be a set of clear and concise statements to formulate the required systems engineering documentation. By following this guidance an agency can expect to produce the following documents:

- Concept of Operation
- System Requirements
- Verification Plan
- Validation Plan

In addition, it is expected that an agency would prepare a procurement plan to provide an appropriate framework for the ASCT acquisition. It is strongly recommended that ASCT systems not be procured using traditional low-bid process, because experience has shown that ASCT systems are complex and require sufficient integration and customization that they cannot be successfully treated as COTS purchases.

A. PREAMBLE

WHY USE SYSTEMS ENGINEERING?

To deploy adaptive control will ultimately require the procurement of equipment and software. Requirements describe to vendors what the agency intends to purchase so that they can propose and bid a solution. Requirements are based on the agency's needs. Systems engineering is a tool that helps the agency articulate its needs.

Federal requirements described in 23 CFR 940 mandate that systems engineering analysis be performed for all ITS systems deployed using Federal funds, and that the level of effort be commensurate with the scale of the project. Also, many ASCT products are considered for proprietary purchase, which, when using Federal funds, requires justification in accordance with 23 CFR 635.411. Proper systems engineering documentation provides such justification. This document will help you respond to those requirements, but more importantly will inform your decision-making process and increase the potential for outcomes that are consistent with and meet the expectations of your system's stakeholders.

PURPOSE OF THIS DOCUMENT

The purpose of this document is to guide the user through the process of developing systems engineering documents for assessment and selection of adaptive signal control technology (ASCT) systems. This document provides a structure within which you can examine your current operation (or the operation you expect to have within the near future), assess whether or not adaptive control is likely to address your issues, and then decide what type of adaptive control will be right for you. Templates are provided for the development of the systems engineering documents that are appropriate for your situation. There are instructions on how to select appropriate answers to questions, how to select statements from the examples that are provided, and what additional information you need to gather and include in the documents. This may lead you to prepare a set of requirements and a specification against which vendors may propose a solution; or it may lead you to identify one system that is particularly suitable for your needs. It may also lead you to the realization that you are not yet prepared or capable of operating an ASCT system.

There are two primary situations in which this document is intended to be used:

- **During the Planning Process:** When you are planning to install an ASCT system, need to define the overall objectives for the system, and need to define it sufficiently clearly that a project can be included in the regional TIP.
- **During the Engineering Process:** Once a project has been programmed and funded, this guidebook may be used to fully define the project and prepare detailed requirements that are suitable for use in the procurement process.

The focus of this guidance is to develop documents that will enable you to procure an ASCT system that is currently available from a vendor. It does not provide sufficient guidance to develop procurement documents for the development of a custom-built ASCT system. That would require much more detailed specification of subsystems, physical equipment and user interfaces.

WHY CONSIDER ASCT?

So you think adaptive signal control may be the solution you are looking for to improve your traffic signal operation. There are many reasons why you may consider that your signal operation could be improved. Let's first look at why you are dissatisfied with the current operation, or expect that it could be operated more efficiently and effectively. You may have existing time-of-day coordination patterns, operating on a fixed cycle length. While these operate satisfactorily much of the time, there are times when the traffic demand varies from that used to develop the timing plans, such as when special events cause the traffic to change, or incidents reduce the available capacity. Even in normal operation, the cycle length in operation at any time tends to be higher than needed for the current traffic volumes, except in the peak of the peaks. This is generally done because there is some variation from day to day in the actual volumes and it is better to err on the side of too long a cycle than too short a cycle. Such variation is often due to: the peaks being of variable duration; daytime traffic varying depending on retail and commercial activity or the popularity of sporting events; numbers of commuters varying with the state of the economy; and the different levels of commuting on normal workdays, national holidays and the numerous partial public holidays.

Traditional coordination plans, whether they are selected by time of day (TOD) or traffic responsive pattern selection (TRPS), also require maintenance. In many communities, traffic patterns change sufficiently within three years that the cost of completely reviewing and updating the timing plans is justified by the improvement in efficiency that results. Although you should not expect adaptive control to be "set and forget", you should expect at least less deterioration in efficiency between timing reviews. Most adaptive control systems operate on top of or in conjunction with many of the most basic low level foundational traffic control settings. Therefore these systems cannot overcome the impact of poorly selected basic timing settings and parameters. It is also particularly important to have up to date, good quality traffic signal control settings and appropriate plans programmed as backup plans that will be invoked if the adaptive control system fails or traffic would be better served by non-adaptive operation.

There are also numerous situations in which coordination of adjacent signals on a fixed cycle length is less satisfactory than some form of free, actuated operation. This may apply where you have several closely spaced intersections with markedly different natural cycle lengths. In such situations, queues from one intersection often affect the adjacent intersection if they are all run on a single, fixed cycle length, and a better alternative is to have certain phases at the lower-cycle length intersections maintain a fixed relationship with the critical intersection while it runs in free, actuated mode.

There is also one situation in which there is no need for coordination; a major isolated intersection. While such an intersection is typically operated in free, actuated mode, relying on gapping and volume-density settings to terminate phases, there are times when a lower, controlled cycle length would be more efficient, particularly if storage lanes are of limited length and queues sometimes overflow into adjacent lanes.

In each of the situations described above, adaptive signal control may offer an improvement over the existing operation. Not all adaptive systems have the same operating philosophies; some are intended as improvements over fixed cycle length coordination; some are applicable to isolated intersection operation; while others extend the actuated, coordinated concept.

There are also various philosophies applied to the architecture of adaptive systems, which will lead to important decisions about the overall management of your traffic signal system. The adaptive operation may be fully integrated into a complete traffic management system. It may be a module within a complete traffic system, or reside in a separate server that is integrated with a complete system. It may be in a separate system that operates completely in parallel with another vendor's traffic management system, or it may be a separate system that takes over control of local signals when it is operating. This process will help you identify your traffic signal system management requirements and then design your system so you can have both the adaptive operation you want and the signal management capabilities you need.

This process will also guide you through the very important steps of system and sub-system verification and system validation. The verification process ensures that the implemented system meets all the requirements you will set, that is, it works as you intended. The validation process measures the extent to which the system achieves the objectives you have set for it. That is, the extent to which transportation is improved by the ASCT in support of your mobility objectives.

While the implementation of ASCT will generally be expected to improve the performance of your traffic system, it is important to bear in mind that, regardless of what system you select, a considerable level of commitment and expertise will be required to realize the full potential of the system. You cannot expect it to be a "set and forget" process. In particular, routine maintenance and periodic or continual performance evaluation will be important elements of the ongoing operation and maintenance of your system.

HOW TO USE THIS DOCUMENT

Systems engineering promotes increased up-front planning and system definition prior to technology identification, selection and implementation. Documenting stakeholder needs and expectations, the way the system is to operate (Concept of Operations), and WHAT the system is to do (the System Requirements) prior to implementation leads to improved system quality.[1] Sections B and C of this document provide you with an overview of the process and a summary of what the final products will look like. Sections D through G will lead you through the development of the documentation by asking questions and requiring you to gather specific information about your agency situation. If you have an existing situation that you would like to improve, answer the questions based on what currently exists. If you know that conditions will change in the near future (e.g., a major new traffic generator is planned to be developed, the road network will change or gradual but significant changes in traffic patterns are expected), then answer the questions based on your expectation of the new situation. The relationships between the various sections of this guidance document are illustrated in Figure 1.

---

[1] CA Systems Engineering Guidebook for Intelligent Transportation Systems Version 3.0 November 2009 (http://www.fhwa.dot.gov/cadiv/segb/)

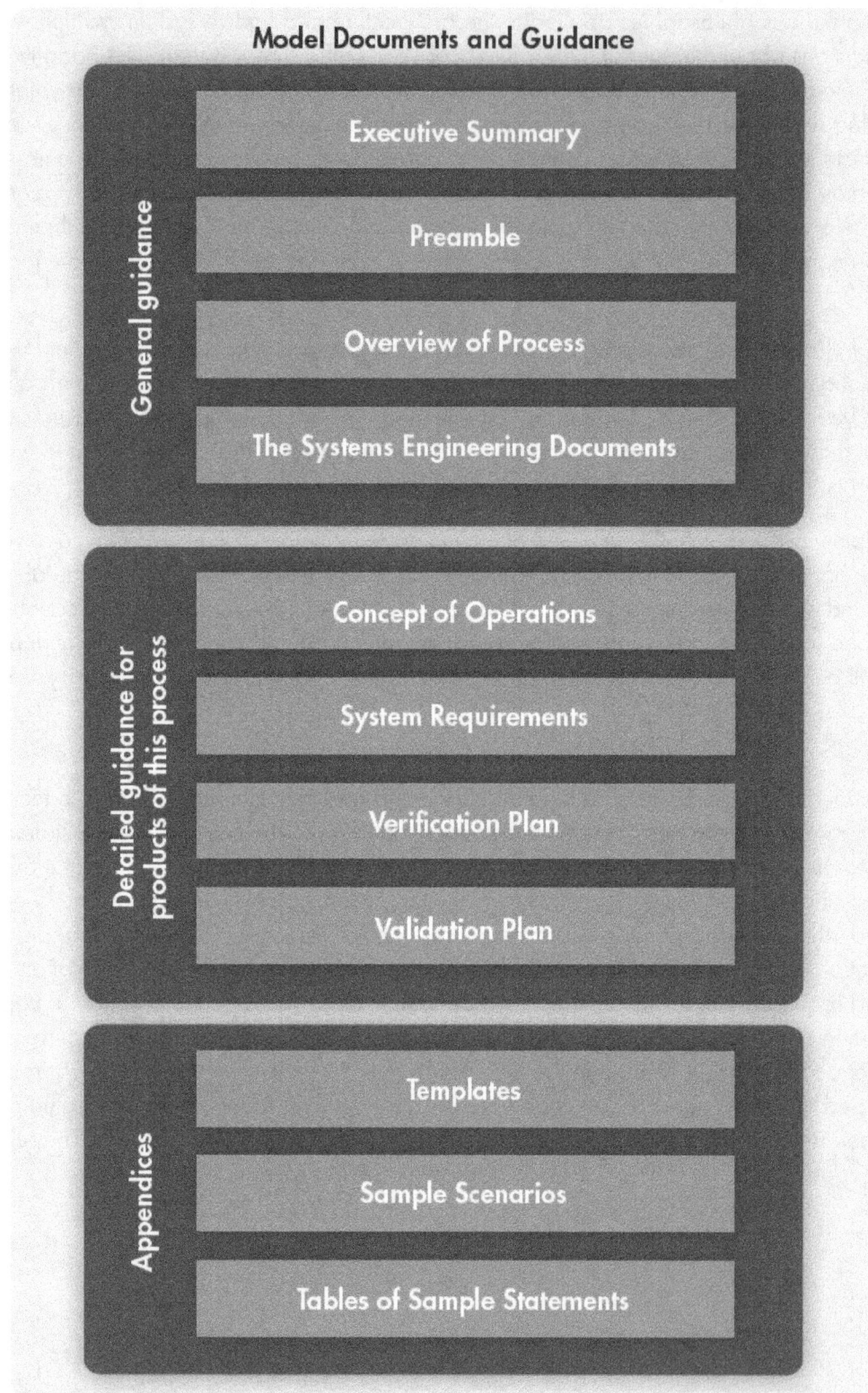

Figure 1. Sections of this Document

You will identify constraints (limitations within which you must work), and you will develop requirements that respond to your documented needs and objectives. **Be careful not to confuse constraints with requirements.** This is a very important distinction, and requires careful differentiation. The requirements that you will select or develop as part of this process should be related entirely to the manner in which you want the system to operate and how you want to interact with the system.

The first phase of this process will help you define what the system *needs to do, not how it will do it*. The second phase of the process will allow you to identify the constraints that may be placed on the system design and/or selection, and will allow you to then decide whether some or all of those constraints can be overcome, or whether they will affect the design. At this point you will know what would be required to overcome a constraint, and what the trade-off will be if you decide to accept the constraint, which will then become a non-functional requirement.

A prime example of a constraint that is often initially (and incorrectly) identified as a requirement is the type of controller that will be used. An early decision to retain your existing controller under any circumstances unnecessarily constrains your considerations and hides the trade-off that you are implicitly making between system capabilities and non-functional constraints.

The third phase of the process will then allow you to decide how the system should achieve your objectives, and represents the design phase. A fourth phase will provide the program to verify and validate that the system you procure actually provides the service detailed in the Concept of Operations and Requirements or in other words *ensures you get what you paid for*.

While it is impossible to definitively estimate the amount of time it will take you to prepare these system engineering documents, and it will vary greatly depending on the size and complexity of the location(s) in which you are considering ASCT, an experienced traffic signal system operator should be able to make a first pass through the Concept of Operations statements and system requirements in one or two days. This first pass will give you a good indication of the likely suitability of ASCT to your situation, and also an indication of the type of ASCT that will be applicable. The length of time required to gather complete data to document your existing situation, needs, deficiencies and proposed operation will be very dependent on the quality and availability of information, the number of stakeholders involved and the complexity of the issues you wish to address.

B. OVERVIEW OF THE PROCESS

Many engineers would like to first answer the question, "Is adaptive control right for my situation?" before proceeding with any analysis. However, there are so many situations in which adaptive operation *may* be better than the existing operation, and such a range of capabilities among the available adaptive systems, that it is almost impossible to answer this question without first developing a clear concept of operation. Only then can you decide whether you should proceed to develop adaptive system requirements. Look at the questions and statements in Figure 2. If one or more of these statements or questions applies to you, then adaptive control may be able to help you, and you should proceed.

Should I Consider ASCT?

I manage a large city, with over 1,000 traffic signals, I'm considering adaptive signal control for some intersections, but how do I determine the right place for adaptive?

I'm a technologist and want to use the latest and greatest. I just heard about adaptive control and it sounds great; I want one! What do I do next to get it?

I have a very old traffic control system and with my recent grant I think I can afford a new system. Is it time to consider adaptive control?

I have tried time of day coordination and even traffic responsive plan selection, but I feel there could be something better. Could adaptive control be a better solution?

I need to improve my network to comply with new air quality standards. Is it time to consider adaptive control?

I been working with my consultant/vendor for many years and they have been telling me about new adaptive traffic control systems that I should consider. What locations would be the best fit for an adaptive control system?

I am getting calls on my couple of my intersections and I cannot solve the cycle/phase issues. Will adaptive control help?

I have a corridor on which I run time of day coordination, but occasionally diverting traffic overwhelms the corridor. Could adaptive control provide a better solution?

The planners are telling me that in the next ten years there will be 50% growth along the main corridor in the city. The current traffic signal system will not handle the traffic based on the current capacity. Is it time to consider an adaptive control?

Figure 2. Should You Consider Adaptive Control?

One statement *not* in Figure 2 is the following: "Our agency desires to eliminate our traffic engineering staff by implementing adaptive control." Adaptive control is often marketed as eliminating the need for signal timing activities, but this is misleading. Observation of adaptive control projects indicates that ASCT improves performance, particularly when resources are limited, but that it still requires expert monitoring and adjustment over time. Even if this expertise is provided by a contractor or vendor, it will be required. Thus, if an agency lacks the resources to operate and maintain a traditional system, then ASCT probably should not be considered until those resources can be provided in a sustainable way.

Having decided to proceed, this process will guide you through the steps that are necessary in order to develop concise and relevant systems engineering documents for your project or situation. *It does not do the engineering for you, that is up to you.* Nor does it cover in detail the procurement, installation and operation of the system. The majority of the guidance covers the many decisions you will need to make in order to clearly define the system that you need. That process is illustrated in Figure 3.

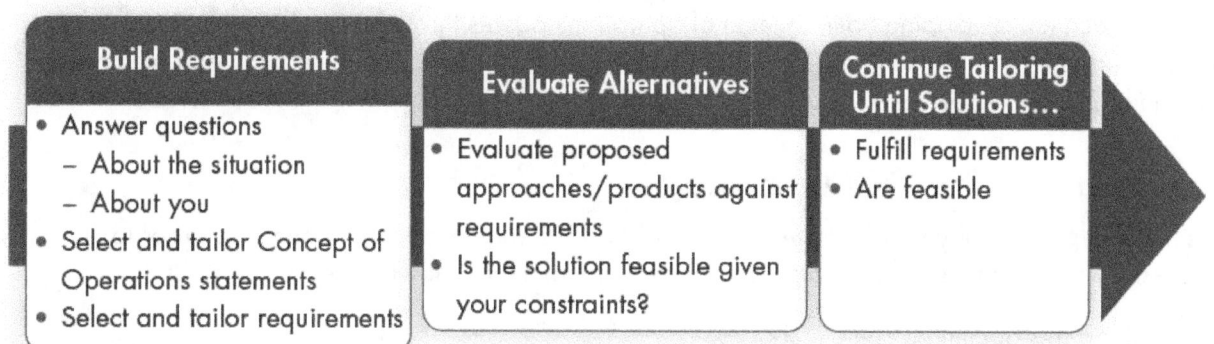

Figure 3. Overview of Systems Engineering Flow Chart for ASCT System Definition

Templates for each system engineering document are included, as a quick reference guide for the structure of each document and the contents of each chapter. More detailed assistance is provided in the remainder of this guidance document.

In order to help you clearly define your needs in terms that will allow you to discriminate between the various adaptive systems that are available, and clearly define capabilities of your existing operation that must be retained with the use of adaptive operation, you will be guided through a series of questions. The answers to some of these questions will guide you to sample statements that may be included in the Concept of Operations. At each step, a description is provided of various situations that will lead you to different decisions. You should look for the statement that most closely describes your situation or aspirations. If you have a unique situation, you will be given guidance on how to describe it and how to frame suitable requirements that would address the situation.

After you go through these steps you will have a good idea of what you wish to achieve, and also the constraints within which you can then make a decision. Once you examine the constraints and decide whether or not any of the constraints can be removed, you will see whether any of the requirements are incompatible with the remaining constraints. This may lead you to reconsider the answers to some of the questions. This will be an iterative process until the remaining constraints are compatible with the functional requirements.

Following completion of the requirements, and *before* a system is acquired or developed, you will prepare a verification plan (to ensure the implemented system satisfies all requirements) and a validation plan (to ensure the operating system satisfies the needs defined in the Concept of Operation).

If the implementation of adaptive control would simply involve the design or selection of a new, stand-alone system that would replace any existing system and functions, the application of systems engineering principles would result in a linear sequence of events. This would involve defining your existing (As-Is) situation; needs, goals and objectives; your desired (To-Be) situation; followed by detailed requirements that will lead to selection and installation of a system.

However, since adaptive control will almost always be implemented in a location where there is an existing system, and the interface between an existing and a proposed adaptive system will be different for almost every combination of vendors and systems, a different approach is necessary in this case. The approach for developing your concept of operations passes through several of the steps multiple times, addressing one element at a time and gradually building up a complete picture of the concept and the high level requirements.

ASSEMBLING YOUR DOCUMENTS

The finished product of your efforts will be several systems engineering documents. In order to successfully prepare these documents, you will need to take the following steps.

1. Read this document completely.
2. Begin to prepare the Concept of Operations. Establish chapters in accordance with the Concept of Operations template. While you are free to format the document to suit your needs, the template follows the outline suggested in ANSI standard G-043-1992. As an alternative, you may simply take the table of sample statements and check those statements you wish to include.
3. Following the instructions in this document, copy and edit relevant statements from the Table of Sample Statements for the Concept of Operations. Depending on how you answer each question in the guidance, select and edit each Concept of Operations statement.
4. Some sections of the Concept of Operations require you to write appropriate text in accordance with the instructions contained in this document.
5. Each statement in the Concept of Operations table has a unique identifier. The needs statements (Chapter 4) each refer to at least one relevant System Requirement that should be considered to support the need statement. Each Concept of Operations statement also contains a reference to the relevant section of this Guidance Document.
6. Begin to prepare the System Requirements. Establish chapters in accordance with the System Requirements template. As an alternative, you may choose to include the System Requirements as an appendix to the Concept of Operations. In this case you may simply take the table of sample statements and check those statements you wish to include.
7. For each need statement used from the Concept of Operations table, identify the System Requirements that are linked. Copy and edit each relevant requirement into the System Requirements document. Note that whenever you pick a "child" requirement, you should also select its "parent" requirement.
8. If you describe needs in the Concept of Operations that are not covered by these sample statements, then you must also create related requirements.
9. Prepare the Verification Plan. Establish chapters in accordance with the Verification Plan template. Note that every System Requirement requires a verification test, which you will need to prepare to suit your situation.
10. Prepare the Validation Plan. Establish chapters in accordance with the Validation Plan template. Note that every need expressed in the Concept of Operations requires a corresponding validation test, which you will need to prepare to suit your situation.

C. THE SYSTEMS ENGINEERING DOCUMENTS

OVERVIEW

The system engineering documents that need to be produced using this process are:

- Concept of Operations
- System Requirements
- Validation Plan
- Verification Plan

A template is provided for each of these documents in Appendix A. In each case, the template is intended as a guide. There is no limitation (upper or lower) on the amount of information included in each document. However, it needs to be sufficient to illustrate that your needs are clearly defined, alternative methods of accommodating those needs have been investigated, there is clear justification for having selected this course of action, there will be no unintended consequences, the design of the system will accommodate constraints within which it must function, the correct operation of the system will be verified and the success of the system in meeting your needs will be validated.

Each of the templates sets out a structure that should be followed, although you are at liberty to modify the chapter titles. The description you provide in each section need not be extensive, provided it covers the elements mentioned in the template. The relationships between the guidance document, templates and the final systems engineering documents you will produce are illustrated in Figure 4.

RESPONSIBILITIES

The responsibility for preparation of these documents rests entirely with the agency. While information may be sought from vendors during the preparation of these documents, a vendor should never be tasked with the responsibility of preparing the systems engineering documents. This would be a clear conflict of interest. However, it is appropriate for agencies to seek support from qualified consultants who are *independent* of vendors. Of the documents described in this guideline, only the verification procedures may have significant input from a vendor, since verification testing requires exercising all elements of the system that are the subject of requirements. The vendor should not prepare the verification plan, and should only prepare procedures if the user manuals are inadequate or not yet completed. Any documentation prepared by a vendor must be thoroughly audited by the agency to confirm its completeness and compliance with the Concept of Operations and the System Requirements.

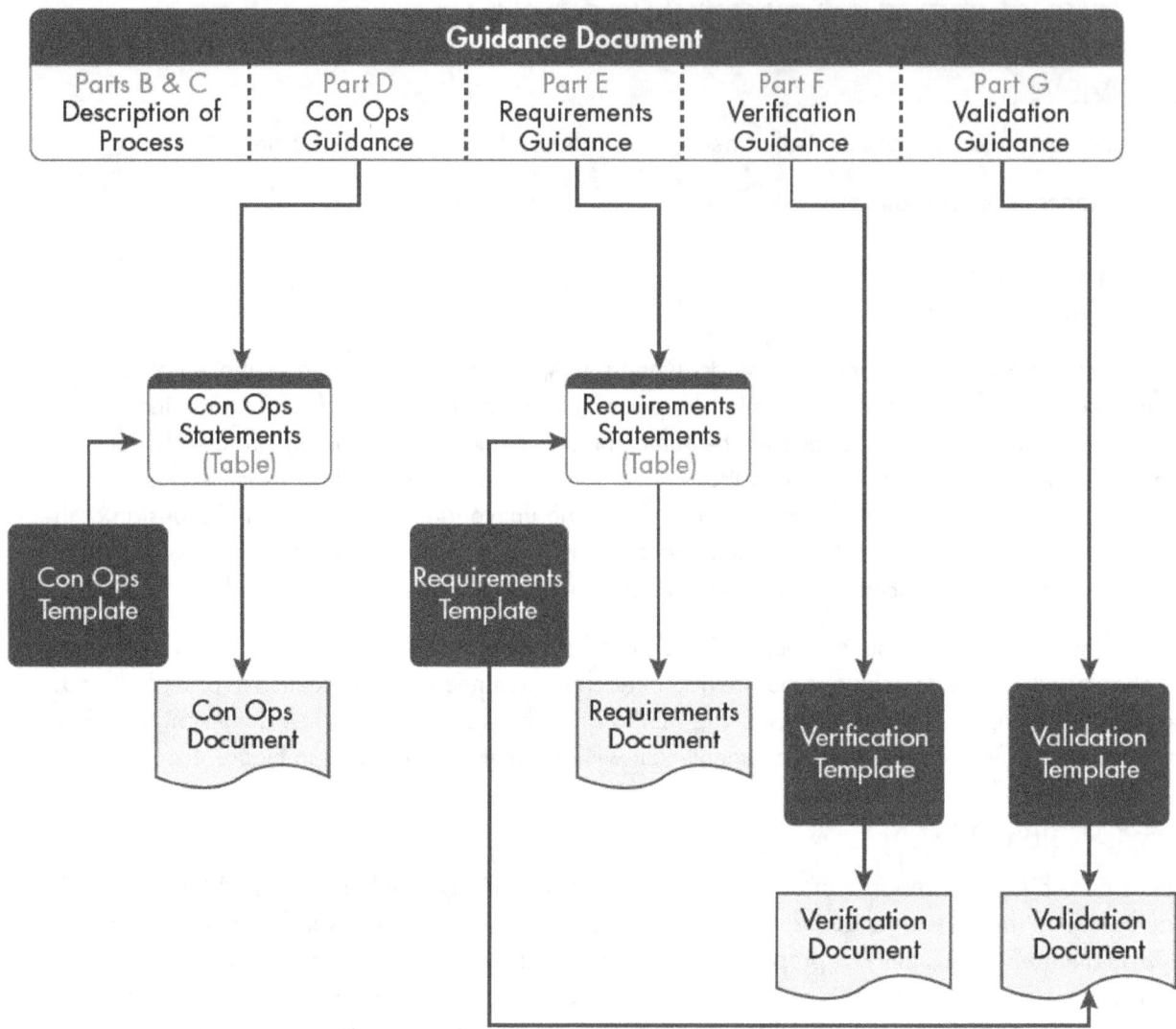

Figure 4. Relationships between this Guide and Final Documents

CONCEPT OF OPERATION

The Concept of Operations is written from the perspective of the system operator, whose story is documented therein. The primary audience for the concept of operations is composed of stakeholders who will share the operation of the system or be directly affected by it. The stakeholders are tasked with describing their needs and objectives succinctly enough to determine what functions the proposed system must be capable of fulfilling. Stakeholders should include system managers, operators and maintenance staff responsible for the system; administrators, decision-makers, elected officials, other non-technical readers may also be included. This document is the key to the success of the project. Every element of all the other documents must be able to be traced to statements of need in the Concept of Operation. Detailed guidance to help you with each chapter is included in Section D.

The Concept of Operations is a non-technical description of HOW the system will be USED. This provides a bridge between the often vague needs that motivated the project to begin with and the specific technical requirements. There are several reasons for developing a Concept of Operations.

- Get stakeholder agreement to describe how the system is to be operated, who is responsible for what, and what the lines of communication are
- Define the high-level system concept to enable evaluation of alternatives
- Define the environment in which the system will operate
- Derive high-level requirements, especially user requirements
- Provide scenarios and criteria to be used for validation (via the **Validation Plan**) of the completed system

When defining the concept and justifying its selection, you must be extremely careful to relate the expected capabilities and potential benefits to your identified deficiencies and limitations

REQUIREMENTS

The target audience for this document is comprised of technical staff, system users, system designers and vendors. This document describes **WHAT needs to be achieved by the system. It specifically should not describe HOW the system will satisfy the needs**. Each of the requirements listed in this document must be linked to a corresponding need described in the Concept of Operations. If you define a requirement that cannot be traced to a statement of need defined in the Concept of Operations, then either the Concept of Operations document must be revised (so its readers will clearly understand why the requirement exists), or the requirement should be deleted. Detailed guidance to help you with each chapter is included in Chapter E of this guidance document.

The types of requirements typically needed to define a proposed system are described in Table 1, although you do not need to differentiate your requirements into these categories.

Table 1. Categorization of Requirements

Requirement Category	Description
Functional requirements	What the system is to do
Performance requirements	How well it is to perform
Non-functional requirements	Under what conditions it will perform
Enabling requirements	What other actions must be taken in order for the system to become fully operational
Constraints	Limitations imposed on the design by agency's policies and practices, such as type of software, type of equipment and external standards
Interface requirements	Definitions of the interfaces between sub-systems or with external systems
Data requirements	Definitions of data flows between sub-systems or with external systems

In all cases, it is essential to describe the functional, performance, non-functional and enabling requirements and the constraints. If the proposed system will interface with another system or requirements are also required at the sub-system level (which may be the case if you are requiring some customization or new modules specifically for your application), then interface requirements and data requirements may also be required.

This document does not define how the system is to be built. This document sets the technical scope of the system to be built. It is the basis for verifying (via the **Verification Plan**) the system and sub-systems when delivered.

VERIFICATION PLAN

The target audience for this document is the same as for the Requirements document. This document describes how the system will be tested to ensure that it meets the requirements. Detailed guidance to help you with each chapter is included in Chapter F.

These documents plan, describe and record the activities which verify that the system being built meets the user needs and scenarios developed in the concept of operations, by fulfilling the requirements described in the requirements documents. Usually, for even moderately complex systems, the following three levels of verification documents are prepared:

- A plan to initially lay out the specific verification effort
- The detailed verification plan that defines a detailed mapping of the requirements to verification cases
- A report on the results of the verification activities

A critical issue is assuring that all requirements are verified by this activity. This is best done by tracing each requirement into a verification case and then into appropriate steps in the verification procedures.

It will not describe detailed procedures for data collection and analysis, although existing procedures defined elsewhere may be referenced. In general, the procedures that will map to the verification plan will be developed by the system developer, supplier or vendor, and approved by the agency, once the system (or appropriate sub-system) has been installed.

Conducting the verification is generally the responsibility of the system developer, supplier or vendor, although the verification tests may be undertaken by the agency. The verification tests should be witnessed by the agency and the results should be audited by the agency to ensure their veracity.

VALIDATION PLAN

The target audience for this document is the same as for the Concept of Operation. This document (or set of documents) describes how the performance of the system will be measured to determine whether or not the needs expressed in the Concept of Operations have been met. It will describe the measures of performance that will be used, the data that will be needed and the type of analysis that will be appropriate. This will include a description of the data and analysis that will need to be available from the proposed system, as well as other data collection and analysis that will be undertaken external to the system. Detailed guidance to help you with each chapter is included in Chapter G.

These documents plan, describe, and record the activities which validate that the system being built meets the user needs and scenarios developed in the Concept of Operations. Usually, for even moderately complex systems, the following three levels of validation documents are prepared:

- A plan to initially lay out the specific validation effort
- The user's/operator's manual and/or a validation plan that defines the detailed operational procedures
- A report on the results of the validation activities

A critical issue is assuring that all user needs and scenarios contained in the Concept of Operations are validated by this activity. This is best done by first tracing each need and scenario into a validation case and then into appropriate steps in the validation procedure.

The Validation Plan may describe detailed procedures for data collection and analysis, although existing procedures defined elsewhere may be referenced. **Validation is the responsibility of the agency, and cannot be delegated to the system supplier or vendor.**

PROCUREMENT PLAN

The method of procurement of an ASCT system has a major impact on the ability of an agency to select the most appropriate system for its situation. 23 CFR 635.411 requires procurement of ITS systems and components to be competitive, unless certain conditions are met. However, competitive procurement does not mandate "low-bid" procurement. In fact, in procurement of complex IT and ITS system, "low-bid" is generally inappropriate. The requirement for competitive procurement is satisfied through an RFP or similar process that allows for careful evaluation of the extent to which each requirement is met and the suitability of the method used to satisfy the requirement. The evaluation process may also include consideration of costs to determine best value for money.

The requirements for a complex system may be classified as mandatory, desirable and optional. All should be considered in the evaluation. Very often, optional can be used to define requirements that will be important at a future date, but need not be included in the package purchased at this time. This ensures that future system expansion and migration paths are not precluded by the initial system design and capabilities.

Steps in the procurement process that may be applicable to your situation are described below.

Request For Information (RFI)

Once your requirements have been drafted, there may be some that you are not sure can be met by any, or a sufficient number, of the available systems. This is a point at which you need to consider whether it will be appropriate to purchase a commercially available system (with or without some customization) or develop a unique system. When this occurs, it is appropriate, and perhaps desirable, to release to vendors a formal request for information (RFI) focused on the specific requirements about which you are uncertain. Response to an RFI should not be a condition of future participation by a vendor in the procurement process.

Use of the RFI will allow you to decide whether or not your requirements can be satisfied by a commercially available system. If not, you can then decide whether to modify your requirements or develop a much more detailed set of requirements and a specification that would be appropriate for development of a unique system.

Industry Review of Requirements

More comprehensive feedback from vendors can be obtained by distributing a draft version of your requirements to vendors. This is appropriate when you have developed requirements that will involve some customization or may include assumptions about appropriate technology. Provided you have included a Concept of Operations and clear statements of need, this provides the opportunity for vendors to contribute in two effective ways.

Vendors will recognize requirements that assume a design that is different from theirs and this gives them the opportunity to redefine the requirement in a manner that does not preclude their system simply because of the wording or structure of the requirement. Vendors will also recognize high cost customization or new development that would be necessary to satisfy a requirement, and this also gives them the opportunity to suggest alternatives that would substantially reduce the procurement cost or streamline the development.

Responses from vendors should be treated as confidential and not shared with other vendors. If a vendor expects a response to be shared, he is less likely to offer advice that he considers proprietary and part of his competitive advantage.

Request For Qualification (RFQ)

A request for qualifications (RFQ) is appropriate when there are mandatory requirements placed on the capabilities and experience of the vendor. This is a means of reducing the agency's risk. It provides the opportunity to assess a vendor's financial stability, their track record in providing support and training, and proof that an advertised system is fully operational and successful. It should NOT be used simply to reduce the workload of the agency in evaluating responses to an RFP by limiting the number of vendors permitted to respond to a detailed specification or RFP. It should be used after requirements have been prepared and the agency is certain that vendors who show satisfactory qualifications will be able to also satisfy the technical requirements.

Request For Proposals (RFP)

A request for proposal (RFP) is the key vehicle for assessing the extent to which an ASCT system will satisfy the agency's requirements. Using a "best value" approach supported by systems engineering provides the most effective use of the system requirements in the procurement process. As illustrated in Figure 5, the requirements are referenced at every step in the process to help guide the selection. While it may contain mandatory, desirable and optional requirements, the vendor should be required to explain how their system satisfies each requirement, and not simply be permitted to provide a Yes/No or Pass/Fail response. Because of the complex nature of ASCT systems, few requirements can simply be submitted to a Pass/Fail test. There are often different ways in which different systems may claim to satisfy a requirement, and not all methods will be equally suitable (or acceptable to an agency) in all situations. Each answer should be evaluated to determine firstly whether or not the answer is accurate, secondly the extent to which it satisfies the requirement, and thirdly as to whether the method of satisfaction is acceptable to the agency.

In addition, the method by which a vendor satisfies a requirement may have other implications, such as the effect on work practices, staff efficiency or requiring additional equipment or software to be efficiently implemented. The method used to compare responses to an RFP should include a means of accommodating the assessment of compliance. The agency must also select an appropriate means of accommodating variation in cost between vendors. Two methods are commonly used:

- A simple cut-off point of acceptable cost, which would allow an agency to obtain the system with the greatest utility within a pre-determined budget ceiling.
- A best-value approach, which would relate some measure of utility to the proposed cost of each competing alternative. This allows the agency to pick the best value for money from two or more closely comparable systems.

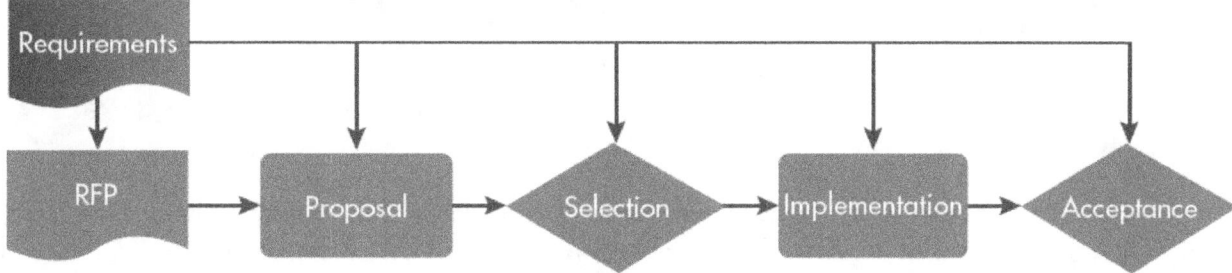

Figure 5. Best Value Supported by SE Analysis

Low-Bid
The use of "low-bid" contracting is rarely a satisfactory method of procuring a complex software system such as ASCT as a complete design-bid-build package. However, it is often possible and appropriate to separate the intelligent portion of a system (such as central software and servers, and local controller software) from the physical components that can be clearly and concisely specified, and construction work that can be undertaken by a qualified contractor.

If this approach is adopted, it is generally appropriate to first use the RFP process to select the ASCT system and vendor. Then the specifications can be completed for the equipment and services to be procured under a low-bid contract with confidence that there will not be compatibility issues between field equipment and the ASCT system.

When purchasing a product, if it is tightly specified then there is limited risk that product will fail to meet the purchaser's expectations. However, when procuring an ASCT system, much of what is being purchased are services, not simply a product. If an agency is restricted to a traditional "low-bid" process, a strategy that builds sound systems engineering into the rigid procurement process (illustrated in Figure 6) would:

- Include System Requirements as part of the special provisions of a standard bid set based on PS&E
- Require submittals at an early stage of the contract to fully demonstrate compliance with the key requirements
- Require a detailed acceptance test plan as an early submittal prior to any construction

It is critical that the compliance with the requirements is demonstrated as early as possible, preferably before any of the project budget is spent.

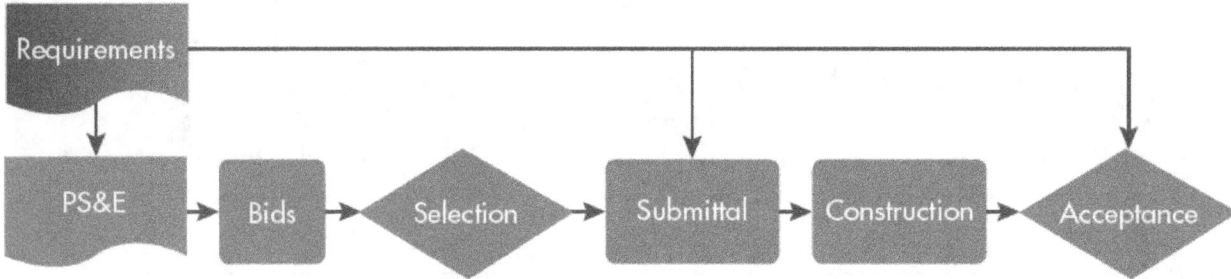

Figure 6. Low-Bid Supported by SE Analysis

Verification Plan
A very important component of the procurement plan is the verification plan. It must be prepared before a formal RFP is issued and, regardless of procurement approach you have selected, before a contract is signed. The verification plan should set out the method by which each requirement will be verified as satisfied, who will undertake verification tests and when within the process each verification test will be conducted. The plan will also include a test and verification matrix that will identify which requirement is verified by each test.

It may be appropriate for compliance with some requirements to be demonstrated prior to selection of a vendor or system, particularly for mandatory requirements that cannot be demonstrated by reference to existing operation with other agencies. For example, if you require an ASCT system to operate with flashing yellow arrow for permissive left turns, and a candidate system has not previously demonstrated such operation successfully, it is appropriate to require this demonstration before finalizing your system selection.

The verification plan should be provided to potential vendors with the RFP. In general, the test procedures are not part of the verification plan at this stage. The procedures can only be written after the ASCT system has been selected and any customization designed. Note that it is usually most appropriate (and cost effective) to schedule tests to occur at various stages during the project, and not leave all testing until after installation is complete.

Market Research Approach
A commonly used but often less successful approach is to begin with market research and employ a low-bid process, illustrated in Figure 7. A detailed set of specifications is prepared based on a list of features and capabilities identified during the market research. It is extremely difficult to avoid specifying unique technology using this approach and the only control over brand choice comes at the submittal stage. The unfortunate experience of many agencies has been that many requirements are only discovered during the acceptance phase at the end of the construction period. This approach has been used for many of the adaptive installations whose final operation has not been satisfactory to the implementing agency.

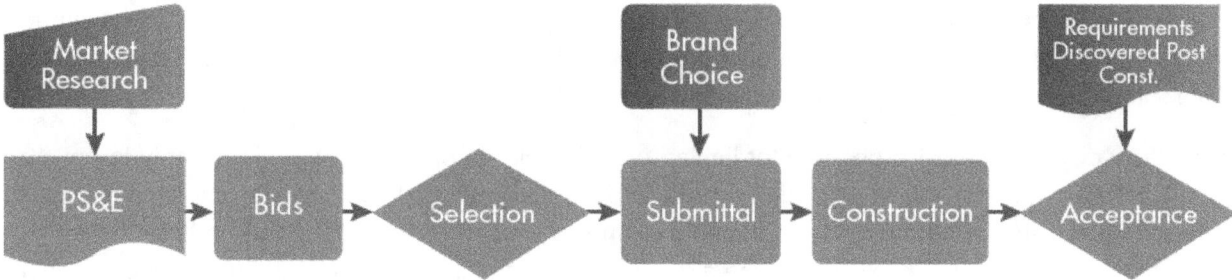

Figure 7. Market Research / Low-Bid Approach

Procurement And Verification References
For further guidance on procurement practices for ASCT and ITS projects, consult the following references.

> NCHRP 560:
> http://onlinepubs.trb.org/onlinepubs/nchrp/nchrp_rpt_560.pdf
>
> Special Experimental Project 14 (SEP 14):
> http://www.fhwa.dot.gov/programadmin/contracts/sep_a.cfm
>
> The Road to Successful ITS Software Acquisition:
> http://www.fhwa.dot.gov/publications/research/operations/its/98036/rdsuccessvol2.pdf

For further guidance on preparation of verification plans, see:

> Systems Engineering for Intelligent Transportation Systems, A Guide for Transportation Professionals
> http://ops.fhwa.dot.gov/publications/seitsguide/section4.htm#s4.7
>
> Systems Engineering Guidebook for ITS
> http://www.fhwa.dot.gov/cadiv/segb/

Sole Source Procurement
In the event that only one system fulfills the requirements, and when federal funds are used, the acquisition process will be governed by 23 CFR 635.411.[2] This regulation governs the acquisition of proprietary materials. In this case, certification from the FHWA is required that one of the following conditions is true:

- The proprietary product uniquely fulfills the requirements imposed on the product. Evaluation of the documents you are creating will provide the necessary justification.
- The proprietary product is required to synchronize with existing systems.

---

[2] Federal Highway Administration, "Construction Program Guide" web page, available at:
http://www.fhwa.dot.gov/construction/cqit/propriet.cfm

A Public Interest Finding (PIF) for proprietary purchases is not required in these cases. A PIF is required when the proprietary product is not the only product that fulfills the requirements. It is recommended that if more than one produce fulfills all requirements, they should be considered competitively.

Proprietary acquisition is also possible for limited experimental application. This might apply to an adaptive system being installed as a pilot project to determine suitability for broader implementation. In this case, you should develop an experimental plan showing what you hope to learn from the experiment that you do not know before the experiment is conducted. Experimental application cannot be used to circumvent the requirement for system engineering.[3] If you have used proprietary acquisition to install a small pilot installation of adaptive operation and you now wish to expand adaptive operation to become a much larger system, a PIF will not be issued simply because you already have a small proprietary system, if other systems may fulfill your requirements.

---

[3] Federal Highway Administration, "Construction Projects Incorporating Experimental Features" web page, available at: http://www.fhwa.dot.gov/programadmin/contracts/expermnt.cfm

D. CONCEPT OF OPERATIONS GUIDANCE

LOGICAL WORKFLOW IN PREPARING THE CONCEPT OF OPERATIONS

While the layout of the Concept of Operations described in this guidance will provide a logical flow for the intended readers, it is generally not prepared in this sequence. As practical traffic engineers, it is generally preferable to describe at an early stage the operational scenarios envisioned by the system operators. After initially describing the limitations of the existing system, you should describe all the situations in which you expect the ASCT system to provide benefit and how you expect the system to operate in each situation. After describing the operational scenarios, you will then be in a position to better describe the specific ASCT system and user needs, the alternative non-adaptive strategies considered and why they were discarded, and the envisioned adaptive system. Then you will be able to revise the operational scenarios so they are consistent with the statements of needs and provide clear examples of the expected operation.

The Concept of Operations will be organized in the following chapters, following the structure recommended in ANSI G-043-1992.

1. Scope
2. Referenced documents
3. User-oriented operational description
4. Operational needs
5. System overview
6. Operational environment
7. Support environment
8. Operational scenarios

Once you have completed the Concept of Operation, use this checklist to confirm that all critical information has been included:

- ✓ Is the reason for developing or procuring the system clearly stated?
- ✓ Are all the stakeholders identified and their anticipated roles described? This should include anyone who will operate, maintain, build, manage, use, or otherwise be affected by the system.
- ✓ Are alternative operational approaches (such as traffic responsive or time of day coordination) described and the selected approach justified?
- ✓ Is the external environment described? Does it include required interfaces to existing systems?
- ✓ Is the support environment described? Does it include maintenance?
- ✓ Is the operational environment described?
- ✓ Are there clear and complete descriptions of normal operational scenarios?
- ✓ Are there clear and complete descriptions of maintenance and failure scenarios?
- ✓ Do the scenarios include the viewpoints of all involved stakeholders? Do they make it clear who is doing what?
- ✓ Are all constraints on the system identified?

The following sections describe how to develop each chapter of the Concept of Operation.

Chapter 1: Scope

1.1 Document Purpose and Scope

The first part of this chapter is a short statement of the purpose and scope of this document. This will briefly describe contents, intention, and audience. Sample statements that may be used in this chapter are contained in the Concept of Operations samples table in Appendix B. These statements should be customized to explicitly apply to your situation. One or two paragraphs will normally suffice.

1.2 Project Purpose and Scope

The second part of this chapter gives a brief overview of the purpose and scope of the system to be built. It includes a high-level description; describes what area will be covered by the project; and identifies which agencies will be involved, either directly or through interfaces. Sample statements that may be used in this chapter are contained in the Concept of Operations Sample Statements table. These statements should be customized to explicitly describe your project. One or two paragraphs will usually suffice. This section should be written late in the process, after the envisioned system has been described. It will be a brief summary to introduce the reader to the proposed system.

1.3 Procurement

The final section of this chapter will be a brief discussion of the proposed procurement process. The method of procurement should be determined early in this process, because it will have an impact on the format and content of the system requirements document.

Chapter 2: Referenced Documents

This chapter is a place to list any supporting documentation and other resources that are useful in understanding the operations of the system. This could include any documentation of current operations and any strategic plans that drive the goals of the system under development. In particular, it should include documents that define the overall goals and objectives of your agency that will be supported by the traffic signal system. This includes local and regional transportation program and policy documents and relevant inter-agency, management and labor agreements and memoranda of understanding. It should reference the regional and local ITS Architecture and include relevant codes and standards, such as ANSI, IEEE, NTCIP, CFR and NEC, and should include relevant codes and standards, such as ANSI, IEEE, NTCIP, CFR and NEC. It should also include references to detailed documentation of any required interfaces to other systems such as a regional traffic conditions map or an Integrated Corridor Management system. However, do not treat this as a bibliography. Only include documents that are referenced directly in the Concept of Operation. Sample statements that may be used in this appendix are contained in the Concept of Operations samples table (Appendix B).

Chapter 3: User-Oriented Operational Description

This is a brief description aimed at non-technical readers who need an understanding of the current system or situation. It should say in only a few words what the existing system is, how it is currently used, what you are currently able to achieve with the system and (most importantly) what you want to do that can't currently be achieved with the system.

3.1 How Does the Existing System Work?

Describe the following aspects of the existing situation in words and figures.

Network Characteristics

Describe the nature of the existing road or network of traffic signals for which you want to consider adaptive operation. Is the capacity of the roads constant, or does it change during peak and off-peak times as a result of parking restrictions and/or reversible lanes? For example, is it an arterial road or a grid, are there several crossing arterials, are there freeway interchanges, is it an isolated intersection or a small group?

Traffic Characteristics

Describe the traffic conditions in the area. For example, is traffic highly directional or balanced; is it heavy only in commuter peaks or also at other times; are conditions relatively predictable or subject to unpredictable fluctuations, due to incidents and diversions; and are there major events that occur frequently at regular or irregular intervals? Include a brief description of pedestrian and public transit characteristics, and how they influence other traffic and the signal operation.

Is the variability limited to a single movement or phase? Does individual phase demand vary considerably from service interval to service interval, not only in when it occurs but also its amplitude?

Has there been an effort to accurately document and study these characteristics and map them to traditional solutions? The results and details of the study should be included.

What are the specific and documented traffic characteristics that are unsolvable with your current systems and approaches?

Signal Grouping

Describe in broad terms the likely grouping of the signals. Are the intersections sufficiently close that they may be coordinated together under some traffic conditions? Are there groups of intersections that are separated by a sufficiently large distance that they will never be coordinated together?

Land Use

Describe the land use in the area. For example, is it residential, commercial, retail, industrial or a combination of these? Are there major, concentrated traffic generators with specific traffic patterns?

Operating Agencies

Describe the agencies involved in the operation of the traffic signal system. This should include the primary agency that operates the signals, other agencies whose signals are under the control of the system, other agencies whose signal systems operate in a coordinated manner with this system (but may not currently be connected to your system); and other agencies (not operating traffic signals) who are affected by the system (such as transit and fire departments) and have some control over the policy, procedures or operation of this system.

Existing Architecture and Infrastructure

Describe the existing system architecture. Provide an appropriate system network block diagram and describe the following elements, as applicable:

- TMC, location of servers, locations of workstations and any associated LAN or WAN
- Intermediate hubs and on-street masters, that are between servers and intersections
- Communications infrastructure, including media, bandwidth and protocols
- Detector locations and technology (e.g., loops, video and other technologies; stop line, advance and mid-block detection zones; number of lanes served by each detector; and any special purpose detectors, such as speed sensors and dual-channel loops with transponder detection capabilities)
- Functions of the existing system currently in use
- When the signals were last retimed and what were the results

Describe the existing Regional ITS Architecture. Show how the existing system fits into the existing Regional ITS Architecture. Include relevant block diagrams that show systems and agencies that are mentioned in subsequent chapters of the Concept of Operation. You may need to revise this section after the subsequent chapters have been completed and after the requirements have been completed.

3.2 What are the Limitations of the Existing System?

At this point, summarize the reasons why the existing operation is considered inadequate and therefore needs to be improved. This may include a brief description of operations or actions the operator would like to be able to implement in order to address various deficiencies or unsatisfactory operations, but cannot with the existing system. The reasons why these actions or operations cannot be implemented may also be mentioned.

3.3 How Should the System be Improved?

Describe in broad terms the general approach to improving the system. Examples of actions or operations that may be desirable include: overcome jurisdictional boundaries that prevent signals from operating together, implement a system that can recognize changes in traffic patterns and react quickly, implement a system that more efficiently accommodates transit vehicles, implement a system that manages queues in critical locations, and implement a system that recognizes differing traffic conditions in various sections of the coordinated network and can make appropriate responses in each section.

At the end of this section, the reader should be able to clearly see the justification for the proposed changes.

Sample statements that may be used in this chapter are contained in the Concept of Operations samples table (Appendix B).

3.4 Statement of Objectives for the Improved System

This section is focused on describing the operational objectives that will be satisfied by the envisioned adaptive operation. This should NOT describe the equipment but rather HOW the equipment will be used. To describe the operational objectives of the system, answer the following question.

What are the operational objectives for the signals to be coordinated?

- Smooth the flow of traffic along coordinated routes
- Maximize the throughput along coordinated routes
- Equitably serve adjacent land uses
- Manage queues, to prevent excessive queuing from reducing efficiency
- Variable, with either a combination of these objectives, or changing objectives at different times
- For a critical isolated intersection, maximize intersection efficiency

In answering this question you should not limit yourself to the current situation. Consider how the objective may change over time, as new development occurs in the area, the number of signals changes and the intensity of the traffic load changes. Consider the period of time over which these changes may occur and compare that with the expected life of the envisioned adaptive system. You may need to select more than one objective as being appropriate for the system.

Smooth Flow
This objective seeks to provide a green band along an arterial road, in one or both directions, with the relationship between the intersections arranged so that once a platoon starts moving it rarely slows or stops. This may involve holding a platoon at one intersection until it can be released and not experience downstream stops. It may also involve operating non-coordinated phases at a high degree of saturation (by using the shortest possible green), within a constraint of preventing or minimizing phase failures and overflow of turn bays with limited length, and with spare time in each cycle generally reverting to the coordinated phases.

Maximize Throughput
This objective seeks to provide a broad green band along an arterial road, in one or both directions, to provide the maximum throughput along the coordinated route without causing unacceptable congestion or delay on the non-coordinated movements. The non-coordinated phases would typically be vehicle-actuated and operated at a high degree of saturation (by using the shortest possible green), within a constraint of preventing or minimizing phase failures and overflow of turn bays with limited length, and with spare time in each cycle generally reverting to the coordinated phases.

Access Equity
Traffic signals are often provided so that major traffic generators along a street can have safe and efficient access to and from the arterial. In these cases, the objective is to equitably serve all traffic movements at each intersection. At the same time, coordination is generally provided along the arterial, but not at the expense of accessibility to local land uses. An example is a suburban retail shopping district that generates significant demand for left-turn and side-street movements, with unpredictable demand characteristics during time periods that are not normally considered when developing traditional coordination plans.

Manage Queues
Where there are closely spaced intersections, such as at a diamond interchange or within a tight grid network, and especially when a short block is fed by movements from various phases, the primary objective is to ensure that queues do not block upstream intersections or movements (such as occurs

when a left turn bay spills over into adjacent lanes, or left turn queues exceed the intersection spacing at a tight diamond interchange). This often requires constraints on cycle lengths and phase lengths to ensure that a large platoon does not enter a short block if it must be stored within that block and wait for a subsequent green phase. It may also involve "gating" a movement, so that a movement is stored at an intersection simply to hold it in a location that has sufficient queuing capacity, even though other movements at the intersection may not require the green time. Multiple phase service may also be an effective tool in management of queues, especially for minor movements where queue overflows can cause problems for major movements.

Variable Objectives

It is often the case that different objectives are appropriate at different times of the day and under different traffic conditions. An arterial road that provides access between a freeway and large residential areas, but also has traffic generators such as retail centers and schools, may require an objective of providing a pipeline maximum throughput during the morning and evening peak periods, but provide access equity during business hours and on weekends, and minimizing stops during other off-peak times.

Maximize Isolated Intersection Efficiency

This objective applies to adaptive control of an isolated intersection. Intersection efficiency may be defined simply in terms of overall delay, or in a more complex objective function that considers stops and other operational parameters.

3.5 Description of Strategies To Be Applied by the Improved System

This section describes the adaptive coordination and control strategies that may be employed to achieve the operational objectives.

Provide a Pipeline

Providing a pipeline along a coordinated route could support the two objectives of minimizing stops along a route and maximizing throughput along the route. The provision of a pipeline along a coordinated route can be achieved by a system based on a common cycle length, and also by a system that provides coordination bands toward and away from a critical intersection without using a common cycle length.

A traditional, cycle-length-based system would use detection to decide on the direction required for coordination (e.g., inbound, outbound, balanced), and it would allow non-coordinated phases to be served and offsets to be set in a fashion that maximizes the width of the pipeline in the desired direction. It will allow phase sequence to be selected to allow the pipeline in both directions to be maximized (e.g., by using lead-lag phasing for left turn phases on the coordinated route).

If the network or arterial road has "resonant" cycle lengths, it will allow those cycle lengths to be identified and vary the cycle length to allow the measured demand to be accommodated within the associated pipeline. Resonant cycle lengths occur on streets with regular signal spacing, where a cycle length that is equal to or 2, 4 or 6 times the travel time between pairs of intersections is a feasible length. These cycle lengths can be described as "resonant" because the relationship between spacing, traffic speed, and cycle length allows platoons in both directions to be served by progression.

If the network or arterial does not have resonant cycle lengths, either because the resonant cycle is infeasible or because of uneven signal spacing, or if the traffic demand requires a higher cycle length to provide enough capacity in the pipeline, the system should allow the cycle length to be adjusted to provide a pipeline with sufficient capacity in one direction.

A non-cycle-length-based system will define the bandwidth of the pipeline to match the phase length of the coordinated phases at the critical intersection within a group. Then the phasing at other intersections will be timed so that green is provided on the coordinated route to accommodate the pipeline.

Distribute Phase Splits
To provide access equity, the demand for all phases will be handled equitably by serving all movements regularly and not providing preferential treatment to coordinated movements to the extent that delays and stops of other movements are significantly increased. To do this a system would be optimizing an objective function that seeks to balance delays or some surrogate measure proportional to delays. Strategies to prevent queue overflow on minor movements may be needed. Typically, a system that distributes green time may do so on the basis of detector occupancy on competing phases or by swapping time between phases based on max-outs and gap-outs.

This strategy may also be applied to a critical isolated intersection in order to maximize intersection efficiency in terms of delays, stops or some composite objective function.

Manage Queues
To manage queues, a system will allow the offsets between intersections to be set in a fashion that allows queues to be cleared at the end of each phase in blocks that are required to store queues during a subsequent phase. It will allow cycle length to be managed to limit queue sizes on designated movements. It will provide a means to control the locations where queues are allowed to form.

This strategy may also be applied to a critical isolated intersection that has limited queue storage on movements that, if they overflow, adversely affect movements in adjacent lanes and therefore reduce intersection efficiency.

Variable Strategies
A system that can accommodate several different coordination strategies will allow each strategy to be selected by appropriate measurements of traffic conditions and will be able to be set up at the operator's discretion to manage queues, maximize throughput, minimize stops along the maximize the coordination pipeline and equitably serve all (or designated) movements.

3.6 Alternative Non-Adaptive Strategies Considered
This section starts with a list and description of the alternative, non-adaptive concepts examined. If you have an existing coordinated system, describe the features of the system that you are not currently using that could potentially be used to improve your operation. Explain why each of these features is not being used. If you have previously tried to implement improvements that have not been successful or are no longer employed for other reasons, describe them along with the reasons they were discontinued.

Traffic Responsive
Many coordinated systems have the capability of using traffic responsive pattern selection (TRPS) to select and engage timing patterns based on measured traffic conditions rather than by a time of day (TOD) schedule. This should be considered in this section.

Carefully consider the potential and limitations of TRPS in your situation. TRPS may be appropriate in situations where you are confident that the traffic conditions will never be outside the range of conditions for which you have prepared and stored timing patterns. The number of alternative plans must be sufficiently small that the pattern selection algorithms can discriminate between the traffic conditions applicable to each pattern. The system's measurement of traffic conditions must be able to include all movements that are affected by the different alternative patterns (or a subset of movements that are reliable indicators of all affected movements).

Complex Coordination Features
There are numerous features available in modern signal controllers and coordinated systems that are often not used. Following is a list of features that may be available within your existing system. You should examine each one and, if it is available, discuss whether or not it is applicable to your situation and whether it would provide the improvement you are seeking with an adaptive system. If you already use a feature within your existing operation operation and that feature is critical to the overall operation and must be retained, particularly when the adaptive system fails, that should be noted in this section. If not, these should be considered for employment as back up strategies in the event the adaptive system fails.

- **Actuated coordination**– Use of vehicle actuation on non-coordinated phases reduces unused or wasted green time and broadens the green band, particularly at intersections with low volume phases. But this approach may cause platoons to be released into the network earlier, resulting in unexpected downstream stops.
- **Multiple (repeat) phase service**– overflow of turning bays can be reduced by operating a turning phase more than once each cycle. This may also provide the opportunity to better coordinate upstream or downstream turning movements and allow better coordination in the non-peak direction by allowing more flexibility in the placement of the through phases in each direction on the coordinated route. Another application of repeat service is when traffic on a side street is light, but the cycle length of the arterial is constrained by the coordination objective. In this case, it is sometimes possible to serve a single vehicle and return to the coordinated phase, then serve another later arrival on the side street and again return to the coordinated phase without adversely impacting the coordination on the arterial. Features related to phase re-service include conditional service or more complex phasing/overlap arrangements. Despite these features, multiple phase service is often difficult and complicated to implement using current signal controllers.
- **Variable phase sequences**– Many agencies maximize the coordination bands by using different phase sequences (particularly leading or lagging left turns on the coordinated route), during different traffic conditions. Different phase sequences may also be used on the side street phases to manage queue lengths on the coordinated route. If your agency has a policy that prevents this operation, that should be made clear in this section.
- **Omit some phases in some plans or at different cycle lengths**– When protected/permitted left turn phasing is used, it is possible to omit the protected phase under some circumstances, such

as at lower cycle lengths or in some coordination plans. This is often used in conjunction with a flashing yellow left turn arrow (FYLTA). This technique provides a wider range of possible cycle lengths in coordination patterns and also allows more efficient free operation at low volumes. If your agency has a policy that prevents this operation, that should be made clear in this section.

- **Detector switching logic to change the function of a detector–** Detector switching can improve intersection efficiency by applying different logic when the controller is in different states, such as to hold the FYLTA on when the through movements gap out early; to extend an overlap when the demand for the overlap movement is greater than the sum of the demands for the underlying phases; or when different demand states exist, such as calling or extending different phases with and without pedestrian demands present.
- **Coordinate different approaches under different circumstances–** The appropriate phase to designate as the coordinated phase is not always the through phase in the peak direction. In some circumstances, such as late-arriving platoons, it may be more appropriate to designate the through phase in the opposite direction.
- **Coordinate turning movements–** In locations with an S-movement, particularly when short block lengths or short turning bays are involved, it may be most appropriate to designate a turning phase as the coordinated phase.
- **Coordinate beginning or end of green–** The coordination strategy you apply to the location should determine whether the beginning or end of green is used as the coordination reference point. Minimizing stops often dictates use of the beginning of green, while maximizing throughput may dictate use of the end of green. In particular, management of queues at closely spaced intersections often requires use of end of green. The selection of the appropriate reference point should be made on an intersection-by-intersection and pattern-by-pattern basis, rather than as a blanket rule.
- **Early release of hold–** It may be appropriate to allow the coordinated phase to gap out early in order to better serve the platoons in the opposite direction during the next cycle, or better serve crossing movements in a network with coordinated cross streets. This may also improve coordination in the peak direction at this intersection and allow the bandwidth of the next cycle to be wider.
- **Hold the position of uncoordinated phases within a cycle–** In a network with coordinated cross streets, it is often desirable to hold the position of a cross street phase in the cycle, rather than start earlier if another phase is skipped or terminates early.
- **Late phase introduction in coordination–** When traffic is light, there may be no demand on a side street phase when it is scheduled to run, and the phase is not introduced. However, if a single vehicle arrives late, rather than making it wait a complete cycle for the next phase, it may be possible to start the phase after its normal introduction time and still return to the coordinated phase at the correct time.
- **Late pedestrian service–** This feature available on some controllers allows the pedestrian walk to be introduced after the phase is already green. This is useful at intersections with high pedestrian volumes, during times when the green time is expected to be longer than the minimum required for the pedestrian service.
- **Stop-in-walk–** This allows for phase green times to be set lower than the minimum time required for pedestrian service. It takes some time away from the start of the next phase but

allows a lower cycle length to be used. It is suitable at locations at which pedestrians are not present every cycle.
- **Dynamic max**– This feature allows the splits to be changed if one or more phases repeatedly run to maximum time and does not gap out. This allows a non-coordinated phase with a short peak that occurs when there is spare time available on the coordinated phases to be accommodated at a shorter cycle length, by using a shorter maximum time for that phase. However, if all phases are running to their maximum time, the feature has no effect. This may also be called critical intersection control (CIC) or adaptive split control (ASC).
- **Specified preemption operation**– This could include specific phase sequences before and after the preemption event. It may also include specific interval timings as well as limited service during the event. The call for preemption could be generated locally or as directed from the central signal systems or other means outside of the traffic signal operational systems.

Sample statements that may be used in this chapter are contained in the Concept of Operations samples table (Appendix B).

Chapter 4: Operational Needs

This chapter lists the needs that will drive the requirements for the system. The system needs will be driven by the answers to the questions on the operational objectives and strategies, desired signal operational features, and the type of adaptive concept you plan to implement. The user needs will be driven by the answers to the questions about user interface, reporting and monitoring and maintenance requirements. Sample statements that may be used in this chapter are contained in the Concept of Operations samples table (Appendix B).

Following is information that will guide your answers to the relevant questions.

4.1 Adaptive Strategies

Before starting to define the desired adaptive operation, it is useful to first review the various operational strategies and coordination techniques that will be considered in the following sections. The existing situation gives a good indication of what the most appropriate type of adaptive operation will be. So, based on the current signal control in the location(s) under consideration, or on the experience with previous attempts to provide coordination, answer the following question. Further explanation is provided below.

In the absence of adaptive control, what is the best operation?

- Coordination around a fixed cycle length
- Actuated, isolated operation
- Actuated operation with one or more intersections slaved from a critical intersection

The following descriptions represent three alternative ways of operating traffic signals. Any traffic signal operation will fall into one of these categories, although there will be many different situations within each category. If you consider that your situation cannot be defined within one of these categories, it is very unlikely that adaptive signal operation will be suitable.

Traditional Traffic Operational Strategies
Coordination Around a Fixed Cycle Length

Fixed cycle length coordination patterns are typically selected by time of day schedules, by traffic responsive pattern selection, or a combination of both. The signal timing within the pattern may be fixed time (the phase splits are identical from cycle to cycle) or semi-actuated (some phases have their duration determined by detecting the presence of vehicles or pedestrians). Coordinated signals using a fixed cycle length cannot be fully actuated, because one or more phases are guaranteed to be green for some part of the cycle, regardless of the presence or absence of vehicles. However, in some systems it may be possible for all phases to be actuated, with part of the coordinated phase being flexible and subject to the presence of vehicles on that phase.

This form of coordination may apply to two or more signals. If your situation currently has coordinated signals, or the proposed situation will benefit from coordination, answer yes to this question. Also, if the spacing and feasible cycle lengths form a natural resonance that provides wide progression bands in both directions, answer yes to this question.

Actuated, Isolated

This situation may apply to one or more signals. In the case of a single, isolated signal, this is normally operated in free or vehicle-actuated mode, with detection on all phases. It relies on detection of vehicle gaps to terminate phases short of the pre-set phase maximum time. If you use alternative maximum times for different times of the day, then adaptive control of the intersection may improve the operation. If analysis of the intersection indicates that different cycle lengths and/or phase splits would be optimal for different times of the day, adaptive control of the intersection may improve the operation.

In the case of two or more signals that have natural cycle lengths that are markedly different, are sufficiently far apart that queues from one intersection do not affect the operation of another, and platoon dispersal is such that vehicles arrive at each intersection in relatively low density platoons, then isolated, vehicle-actuated control is most likely the most efficient operation. If all three of these conditions apply, then consider each intersection in isolation as described in the preceding paragraph. If not all three apply, then coordinated, actuated operation may provide a more efficient alternative.

Actuated, Linked

There are many situations in which a critical intersection is close to one or more minor intersections and the key operating objective is to ensure that queuing from the minor intersections does not adversely impact the operation of the critical intersection. This is often the case when there are frontage roads or other streets close to a freeway interchange that is operated as a single intersection. In such cases, it is desirable to operate the minor intersections as slaves to the critical intersection. While traditionally this may be achieved using hard-wired holds and releases, this can become unwieldy or impossible if several signals are slaved from the one master controller. Some adaptive systems provide the ability to run the greens at minor intersections in a manner that ensures progression to and from the critical intersection even when it is operating in free, actuated mode.

Adaptive Coordination and Control Techniques
Common Cycle Length

There are several types of adaptive systems that use a common cycle length to coordinate groups of signals. Each of these may be implemented with vehicle actuation on some or all phases, or no vehicle actuation. An adaptive system may be able to implement one or more of the following approaches, and the approach may be able to be selected on the basis of measured traffic conditions or a time-based scheduler.

- Use a cycle length that has been previously determined or implemented by another system, and optimize the phase splits and offsets within that cycle length. A previously determined cycle length may be appropriate when the network topology is such that there are one or more "resonant" cycle lengths that are multiples or sub-multiples of the travel time between important intersections.
- Calculate an appropriate cycle length based on traffic demand, and then calculate splits and offsets that are also suitable for the traffic demand. This may be used to select from several pre-determined cycle lengths that are suitable for the network topology.
- Calculate an appropriate green band width and a related set of offsets to accommodate traffic on a coordinated route, then select a cycle length and phase splits that are compatible with that green band.

Systems that use a common cycle length are referred to in these documents as "sequence-based," because when demand is present, each intersection displays the currently permitted phases in the same sequence from cycle to cycle.

Isolated Actuated Adaptive Operation

In this type of operation, an adaptive system may calculate an appropriate cycle length and splits for a signal, or calculate appropriate splits within a specified cycle length and apply them to an isolated signal that is not coordinated with any other signal. In the latter case, the cycle length may be specified according to a schedule. This operation may also be characterized as "sequence-based."

Actuated Linked Operation

This type of adaptive operation may or may not use a repetitive cycle length, and may allow a flexible sequence of phases. If the sequence of phases is flexible at all intersections, there is no cycle length, and this operation is referred to in these documents as non-sequence-based. At the critical intersection, the sequence of phases and length of time provided for each phase will be determined by an objective function or other logic seeking to minimize delays and/or queue lengths in some fashion. This function may be definable by the operator. At the non-critical intersections, the green on coordinated routes will be displayed to provide progression of platoons traveling toward and/or away from the critical intersection. This may be done in a manner that optimizes some objective function such as number of arrivals during the green phase.

Phase-based Operation

Unlike the sequence-based and non-sequence-based operation described above, phase-based operation is generally restricted to a very small number of signals, typically one or two minor intersections close to one critical intersection. In this operation, phases at the minor intersections are generally timed in relation to specific phases at the critical intersection, while the critical intersection may operate in either vehicle-actuated mode or at a specific cycle length. This approach to coordination is often suitable for a freeway interchange with nearby frontage road intersections.

4.2 Network Characteristics
What is the size of the network that needs to operate adaptively, both initially and in the future?

If you are considering adaptive control for a group of signals, does this group include all the signals that you are likely to ever want to operate adaptively, or are you considering this a demonstration that, if successful, may be expanded to include other groups of signals? Answer the following questions:

- How many intersections in total need to operate in adaptive mode?
- Will the signals be divided into groups that will be expected to operate together in a coordinated fashion all or some of the time? If so, how many signals will be assigned in each group, and what will be the largest number of signals in one group?
- Will the number of signals in a group need to be flexible, or will it be constant?
- If the groups are widely dispersed, what is the distance between them? When the groups are dispersed, will they operate independently at all times?
- Describe the expected interactions and relationships between the groups.

Number of Signals
This is the total number of signals that may be operated under adaptive control, in all locations at which adaptive control may be considered. Depending on the size and responsibilities of your agency, this may include signals in different localities separated by hundreds of miles, or may simply be a small group of intersections on one arterial in a small community. The type of adaptive control you may have in mind or finally specify may differ from one location to another, and this will be considered in a subsequent step.

Size of Groups
Determine the number of intersections that may need to be controlled together as a logical group. Coordinated groups of signals may naturally be separated by distances that are sufficiently long that platoons tend to disperse and need to be regrouped to be handled effectively. The nature of a continuing arterial road may change to the extent that efficient operation in one section is quite different from efficient operation in another, such as changing from a four lane undivided road to a six lane divided road. Consideration of these factors will define the maximum number of signals that need to be controlled in one logical group under adaptive control.

Flexible Groups
In typical TOD coordinated systems, the grouping of signals often varies by time of day. For example, an arterial road may have the same cross-section for its entire length, but have several different and distinct sections, each with different traffic characteristics. This may be illustrated by Ygnacio Valley Road, Walnut Creek, CA. At its western end, it connects to two freeways adjacent to the downtown and has several crossing arterials. Its central section has one crossing arterial and serves schools, a major medical center and small retail center. Its eastern section has one crossing arterial and serves a large business park adjacent to that intersection, with access to the business park via several signalized intersections on each of the two arterials. At different times during the peak periods, the heaviest movements may be westbound, eastbound, or relatively balanced, while the total volume is similar along its entire length. During business hours, the desired cycle length is different in each section. As a result, when the TOD signal timing plans are prepared, the signals may be grouped into one, two or three separate coordinated groups, depending on the desired cycle length, the volume of traffic in

different sections, and the lengths of queues that appear at the boundaries of the groups if they are not synchronized.

Using the same logic, you may expect that the logical group you have defined may not be constant under all circumstances. Would you expect an adaptive system to vary the composition of the logical control groups under any of the following circumstances?

- If the system is cycle-based and the selected cycle lengths of adjacent groups are close, should they be forced to operate at the same cycle length?
- If queues at the boundaries between two groups may become long and interfere with the operation of the adjacent group, should they be forced to operate as one group?
- If the volume traveling between two groups exceeds some threshold, should they be forced to operate as one group?

Another example of the intersection grouping not being constant occurs when an arterial road or heavily used route lies within a grid network. During peak times, it may be common to operate the arterial road and the grid streets independently. However, during evenings and on weekends when traffic is lighter, the signal timing may be determined by the street geometry rather than the traffic volumes, and the most efficient operation may be to combine all intersections into one coordinated group.

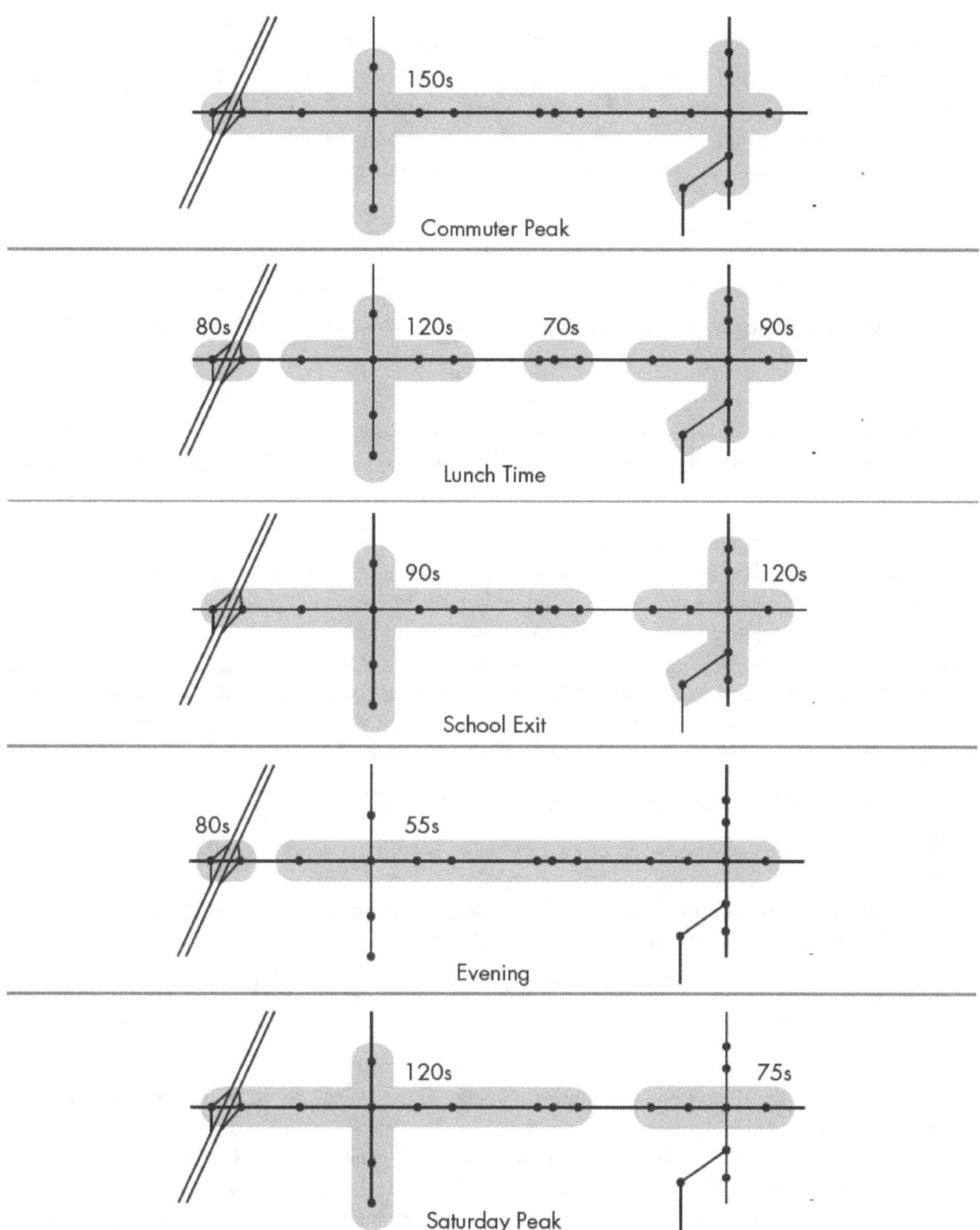

Figure 8. Example of Flexible Grouping

Separation of Groups
If you have defined logical groups that are in widely separated locations, are there any circumstances under which they may need to be operated in a coordinated fashion? This may include synchronizing them under unusual circumstances, such as when there is major diversion of traffic due to an incident elsewhere, or during special events. At this time, do not consider whether they should be under the control of the same system; this will be considered in a later section.

Relationship of Groups to Other Groups or Facilities
Are there multiple groups of signals that impact (and interact with) each other? In some situations the groups may be parallel or perpendicular or crossing.

Will a group be impacted by major sources or sinks such as freeway interchanges, parking facilities, shopping malls, rail facilities, and event venues?

Jurisdictional Relationships
Describe the relationship of your agencies to those directly involved or immediately adjacent to the proposed adaptive system. Who are they and what are their expectations for the project and what can they bring to the table.

4.3 Institutional and System Boundaries

Often, the signals operated by one agency are adjacent to signals operated by another agency and they are timed to operate in a similar and compatible fashion. This may apply to groups of signals along one arterial road (e.g., crossing boundaries between adjacent cities), to groups of signals on crossing arterials (e.g., a city arterial crossing a state highway), to groups of signals in a grid network adjacent to an arterial road (e.g., a downtown area straddling a state highway). It may also apply to small groups of signals of one agency isolated within a larger group operated by another agency, such as interchange ramp signals on an arterial road. Answer the following question.

How will borders with other systems and institutions be handled?

- Make neighboring system part of the operation, but remain separate system
- Make neighboring system part of the system
- Make the adaptive system respond to the operation of the neighboring system
- Allow stratified levels of authority and control

These situations are very similar to the considerations of the relationship between an adaptive system and crossing arterials.

Part of the Operation
In this arrangement, the signals operating in the adjacent system would not be part of the adaptive system. However, they would operate in a manner consistent with the adaptive system, so that coordination is not broken at the boundary between the two systems. This could be achieved by providing communication between the two systems, and have the adaptive system transmit key operational information such as cycle length, time clock synchronization data and offset data, so the non-adaptive system could adjust its operation in response to changes in the adaptive system's operation.

Part of the System
In this arrangement, the signals operating in the adjacent system would become part of the adaptive system.

Constrained Adaptive Operation
In this arrangement, the adaptive signals would operate within constraints imposed by the adjacent system, so that coordination is not broken at the boundary between the two systems. For example, the adaptive system may operate on the same cycle length that is in operation in the adjacent system, and optimize its phase splits and offsets along its coordinated route. This could be achieved by receiving data from the other system(s) on cycle length, synch points and offset requirements. It may also be achieved by sensing the traffic conditions within the other system(s) and adjusting the adaptive operation to accommodate that traffic.

Crossing Arterial Coordination
Often an arterial road that is coordinated will be also coordinated with other arterials and streets that cross it or it may be part of a grid network whose signal timing is determined by that of the arterial. If the route for which you are considering arterial operation is currently also coordinated with cross streets or an adjacent grid, you need to consider whether that relationship should continue with the adaptive system. Answer the following question.

If there are crossing arterials, how will compatibility be maintained?

- Make them part of the operation, but remain separate systems
- Make them part of the system
- Constrain the adaptive system to respond to the operation of the crossing arterial

Part of the Operation
In this arrangement, the signals operating on the crossing arterials would not be part of the adaptive system. However, they would operate in a manner consistent with the adaptive system. This could be achieved by providing communication between the two systems and having the adaptive system transmit key operational information such as cycle length, synch point and offset data, so the non-adaptive system could adjust its operation in response to changes in the adaptive system's operation.

Part of the System
In this arrangement, the signals operating on the crossing arterials would be included within the adaptive system.

Constrained Adaptive Operation
In this arrangement, the adaptive signals would operate within constraints imposed by the systems controlling the crossing arterials. For example, the adaptive system may operate on the same cycle length that is in operation on the crossing arterials and optimize its phase splits and offsets along its coordinated route. This could be achieved by receiving data from the other system(s) on cycle length, synch point and offset requirements. It may also be achieved by sensing the traffic conditions within the other system(s) and adjusting the adaptive operation to accommodate that traffic.

Adaptive Coordination Strategies

Part of the Operation

The adaptive system optimizes its operation on an arterial in accordance with its own objective functions. It then communicates key data to the adjacent systems operating other intersections, and those systems adjust their operation to maintain consistency with the adaptive system. This would typically involve sending cycle length, synch point and offset data to the other systems.

Part of the System

In this arrangement, there would be one integrated adaptive system controlling all the arterials.

Constrain Adaptive Operation

There are two ways in which an adaptive system could operate or be constrained to operate in concert with an adjacent system over which the adaptive system does not exert influence.

- Use adaptive control that does not vary cycle length, but accepts cycle length as an input. The adaptive system will then adjust splits and offsets at the intersections along its coordinated route, based on the measured traffic conditions.
- Detect platoons leaving the adjacent system and approaching the adaptive system. Adjust the adaptive system's operation to accommodate the arrival of those platoons.

4.4 Security

There will be staff in different sections of your agency, and in different agencies, who will need to have access to the system to perform different functions. For example, an operator will need to be able to modify parameters to be able to set up and fine tune the operation of the system. Other operations staff will need to be able to generate reports of system performance, but not need to modify parameters. Operations staff in other agencies may need to monitor the traffic conditions within the adaptive system and could be given permission to modify its operation under certain circumstances. To achieve this arrangement, the system needs to have assignable security levels and jurisdiction rights that allow an administrator to assign appropriate privileges to each potential user.

4.5 Queuing Interactions

The presence of queues within the area proposed for adaptive control, or the risk of adaptive operation causing queues that affect other elements of the transportation system, are very important to the required capabilities of the envisioned system. Answer the following question.

Are some queues outside the control of the envisioned system?

- Will queues back into the system from downstream congestion?
- Do queues form within the system from traffic generators rather than due to signal operation?
- Will queues that propagate outside the system be unacceptable?
- Is it important to control where queues are stored within the system boundaries?
- Is there a need to flush queues through the system?

Will queues back into the system from downstream congestion?

A common situation on an arterial road that is serving freeway on-ramps is for congestion on the freeway or queuing from a ramp meter signal to cause a queue that may block an intersection within the adaptive group. In these cases it is necessary to detect the presence of a queue before it causes an adverse impact on the adaptive system and adjust the system's operation to prevent the queue extending into the system.

Do queues form within the system from traffic generators rather than due to signal operation?
If a queue builds up at a location within the network that cannot be controlled by the adaptive system, such as at the entrance to a parking garage during a major event, it is generally desirable to detect the queue and modify the system's operation to prevent the queue growing to an extent that it compromises the system's operation.

Will queues that propagate outside the system be unacceptable?
At locations such as an intersection with a freeway off-ramp, excessive queuing on the off-ramp may have consequences that outweigh the objective of most efficiently coordinating an arterial. In such cases, it would be desirable to detect the presence of queuing at entry points to the adaptive system and modify its operation to prevent the queue from growing.

Is it important to control where queues are stored within the system boundaries?
If there are short blocks between adjacent signals, such as at freeway interchanges, it is often necessary to ensure that some approaches are always clear of queues at the end of each phase. To do this often requires the green time at one intersection to be controlled so that queues form at that intersection and are only released downstream when they can clear the next intersection. This requires the adaptive system to have the ability to control the green time and the offsets so queues are stored at the desired intersection and released at the desired time. Queues may also be prevented from overflowing by use of multiple phase service. If preventing queue overflows from specific movements requires more frequent service than is feasible for all movements, then multiple phase service is needed.

Is there a need to flush queues through the system?
This consideration is related to the previous, in that the system needs to be able to ensure that once a queue is released it is not stopped until it passes a designated intersection.

4.6 Pedestrians
How should pedestrians be accommodated by the adaptive system?

- Are pedestrians central to the operation?
- Special features (e.g., pedestrian early start, variable lane/movement assignment, variable turn restrictions, programmed pedestrian recall)
- What adaptive response is required?
- What current techniques must be retained?

This question will help you decide to what extent the accommodation of pedestrians by the system will impact your selection of an adaptive system. If your system has little expectation of pedestrian activity, then accommodating or optimizing the operation with the pedestrians intervals included is not needed. Rare pedestrian activity can be accommodated by the local controller, even if it must override adaptive operation in these cases, with no overall detrimental effect. But if pedestrian activity is frequent during critical periods, the adaptive control should accommodate the pedestrians and optimize the operation around them. If pedestrians are accommodated using more complex features, you will need to carefully define the manner in which pedestrians must be accommodated by the adaptive system.

Examples of more complex features include: changing the phasing and/or allowable movements on an approach to a signalized intersection when the presence of pedestrians causes unacceptable queuing and coordinating phases that serve pedestrians at adjacent, closely spaced intersections.

Do you need the adaptive system to:

- Disregard pedestrians when calculating the adaptive signal timing, but serve rare pedestrians locally, ignoring the adaptive timing
- Incorporate accommodation for occasional pedestrians in the adaptive signal timing
- Always allow the full pedestrian time in each phase
- Allow custom pedestrian features
- Some or all of the above, depending on the situation

In this section, give a brief overview of pedestrian operational needs. The details will be examined as the requirements are defined. The Concept of Operations table contains sample statements that may be used.

4.7 Non-Adaptive Situations

From time to time it is possible that traffic conditions would fall outside the range of conditions that an adaptive system can accommodate. When these conditions occur, one of the fallback non-adaptive modes of operation should be employed. There are various ways in which this may occur.

Are there situations that may lead you to sometimes require non-adaptive control?
Do you want the system to:

- Deal with it adaptively and automatically?
- Go to non-adaptive control during the presence of a defined condition (such as exceeding volume and occupancy criteria on specified detectors)?
- Operate non-adaptively according to a user-defined schedule?
- Operate non-adaptively during special events and diversion around incidents?

Adaptively and Automatically

In this situation, the adaptive system would be capable of adopting a wide range of operations covering the full range of expected traffic conditions. The system's objectives would seamlessly move from one to another depending on the measure traffic conditions, and its operation would change to suit those conditions without manual intervention.

Detect Conditions

In this situation, the adaptive system would recognize that the conditions are outside its operating range and automatically choose a non-adaptive mode of operation while the conditions remain in this state. It would automatically revert to adaptive operation when conditions returned to the acceptable range. An example of this situation is diversion along a corridor arterial around a freeway incident. The adaptive system would detect that the incident has occurred, or detect the diverting traffic itself.

Schedule Operation

This situation would involve using a time-of-day scheduler to direct the system to operate in a non-adaptive mode. This may be appropriate when there are predictable conditions to which the adaptive system may not be able to react, or may react more slowly than necessary. An example of such a situation is the change of shift at a large factory, or end of classes at a large school, both of which often result in a sudden and predictable increase in traffic volume exiting a parking lot.

Operator Override
In this situation, an operator would manually force the system operation to a non-adaptive mode. Typically, this would require operators to use stored plans that had been previously developed, or manually modify operation in real time in response to observed conditions.

4.8 System Responsiveness

Depending on the nature of the traffic within the system's area, you may or may not want the adaptive system to continually adjust to small changes in the traffic demands. Answer the following question.

How responsive do you want the system to be to:

- Small shifts in demand?
- Large shifts in demand?

Small Shifts in Demand
Depending on the nature of the traffic within the system's area, you may or may not want the adaptive system to continually adjust to small changes in the traffic demands. To some extent this reflects the different capabilities of the various systems, but also will depend on the type of road network that will be coordinated. For example, a system that maintains a constant cycle length can efficiently vary splits from cycle to cycle. A system that constantly adjusts cycle length can also efficiently react to small changes in demand. However, if the arterial road or network has one or more "resonant" cycle lengths, small changes in demand will not require changes in the underlying cycle length.

Large Shifts in Demand
Large shifts in demand should be detected and accommodated by all adaptive systems. However, the nature of the traffic conditions will determine whether the system will need to react quickly to sudden large changes, or gradually. If large shifts in demand occur relatively gradually over a period of time, such as the buildup and decay of most peak periods, then that will define the rate at which the adaptive system should be expected to react. However, if the system operates in an environment in which the level of demand can vary dramatically without a predictable schedule, such as emptying a parking lot at the end of a large sporting event or accommodating traffic diverted from a freeway as the result of a peak hour incident, then you may require the system to detect the change and react swiftly. If the change in demand level is likely to be observed by an operator, then this rapid response may not be required.

Response Tme
It may also be appropriate to have different response times required for adaptive solutions and non-adaptive solutions to measured conditions. For example, if a large shift in demand is detected, and the condition is outside the range of acceptable conditions, it may be desirable to immediately implement the non-adaptive solution. However, within the acceptable range, a slower response may be more acceptable in order to prevent oscillations or instability in the system's operation.

You may wish to define different rates of response to different conditions. Several examples are included in the Concept of Operations sample statements and include such situations as: the normal rise and fall of volume around peak periods; rapid increases in volume during diversion from an incident, change of shift at a factory, or end of the school day at a high school; the detection of spillback from a left turn queue or a ramp metering signal; the prediction of queue spillback through recognition of repeated phase failures.

4.9 Complex Coordination and Controller Features

Answer each of the following questions to determine what advanced controller functions need to be maintained during adaptive operation. Each of these items is described in more detail below.

- Do you need multiple phase service?
- Do you have multiple overlap phases?
- Do you permit variable phase sequences?
- Do you omit some phases in some plans or at different cycle lengths?
- Do you use detector switching logic to change the function of a detector?
- Do you have non-standard features? If so, identify them.
- Do you coordinate different approaches under different circumstances?
- Do you coordinate turning movements?
- Do you require early release of hold?
- Do you require the ability to hold the position of uncoordinated phases within a cycle?
- Do you allow late phase introduction in coordination?
- Do you use protected/permissive phasing?
- Do you require Flashing Yellow Arrow protected/permissive and permissive only left turn service?
- Do you require movement restrictions by volume or time-of-day? For example a left turn restriction (possibly phase omit) by TOD?
- Do you have special requirements for one lane of an approach?
- Do you require transit queue jump operation by TOD or adaptive?
- Do you require specific phases and sequences to occur following a pre-emption event?

Multiple Phase Service

Multiple phase service may be required for a left turn phase that has a short turning bay that would overflow if the turning phase was simply leading or lagging the opposing through movement. In this case it is efficient to run the left turn phase both before and after the opposing through phase. This may operate with both protected only left turns and protected/permissive left turns.

When traffic on a side street is light, but the cycle length of the arterial is constrained by the coordination objective, it is sometimes possible to serve a single vehicle and return to the coordinated phase, then serve another later arrival on the side street and again return to the coordinated phase without adversely impacting the coordination on the arterial.

This section applies to situations where the multiple phase service is provided by the signal controller and must be tolerated and accommodated by the adaptive system.

Do you have multiple overlap phases?

An intersection with complex channelization may have multiple overlap phases to accommodate heavy turning movements, rather than simply relying on right turn on red to minimize delays to right turning traffic.

Do you permit variable phase sequences?
Many agencies maximize the coordination bands by using different phase sequences (particularly leading or lagging left turns on the coordinated route) during different traffic conditions. Different phase sequences may also be used on the side street phases to manage queue lengths on the coordinated route. This would also include actuated extended pedestrian phase intervals.

Do you omit some phases in some plans or at different cycle lengths?
Do you use protected/permitted left turn phasing, and omit (exclude) the protected phase under some circumstances, such as at lower cycle lengths or in some coordination plans?

Do you use detector switching logic to change the function of a detector?
Do you use logic to change the phase call and extend functions of a detector, based on the state of the signals and the state of demands for other phases? If so, describe the logic.

Do you have non-standard features?
Are there special operating features that you need to retain during adaptive operation that have been customized for your situation? For example, an intersection with a wide central median may have pedestrians cross in two stages, or may operate the pedestrian crosswalks on each side of the median independently, and may overlap with vehicle phases that would normally be in conflict in an 8-phase operation. Describe any such features.

Do you coordinate different approaches under different circumstances?
Do you sometimes coordinate the through movements on one road, and at other times coordinate the movements on the crossing street at the same intersection?

Do you coordinate turning movements?
Do you sometimes coordinate a turning movement rather than a through movement on an arterial road? There are several relevant examples of this situation. If you have the coordinated route with the heavy movements turning left or right at an intersection, it may be desirable for the left turn phase at that intersection to be the coordinated phase. The coordinated route through a network or in one section of an arterial may involve a heavy S-movement, which may require the coordinated phases at two adjacent intersections to be turning phases, or one turning phase and one overlap.

Do you allocate unused green time to a non-coordinated phase?
At a critical intersection on an arterial road, or at the crossing point of two arterials or two coordinated routes in a network, it is sometimes desirable to allow unused green time from turning phases to be added to through movements that are not the coordinated phase.

Do you need the ability to choose which phase to allocate unused time to? For example, you may wish to add unused time to the next phase, to a later uncoordinated phase, or to the beginning of the next coordinated phase.

Do you have special requirements for one lane of an approach?
Do you have downstream movements that result in queuing in one lane of an approach while the adjacent lanes are not congested? A typical situation of this nature occurs when there is a heavy right turn onto a freeway ramp. The queue for the right turn may extend upstream through several intersections in one lane only. In such a situation, there may be no benefit in extending the green for

that movement, so the adaptive system would need to be able to discriminate between lanes and react differently to demand in each lane.

Early Release of Hold
It is sometimes desirable to allow a coordinated phase to terminate early, once a platoon has passed through the intersection, so that unused green time can be added to the beginning of the next coordinated band. This often improves the coordination in the non-peak direction.

Hold the Position of Uncoordinated Phases
At times it is desirable that if a phase has no demand, the preceding phase continues green rather than immediately skipping to the next phase. This is known by various terms such as sliding yield window, yield by phase, multiband permissive and false green.

Late Phase Introduction
When the cycle length is determined by the network geometry and the traffic volume on side streets is light, the "pipeline" along the arterial route is often not fully utilized and the side streets often gap out and return to the coordinated phase early. When there is no pedestrian call on the side street and no vehicle call at the yield point of the coordinated phase, the side street phase is typically skipped. However, some systems allow the side street phase to be introduced later in the cycle if there is still no pedestrian call and a late-arriving vehicle could be served without forcing the intersection out of coordination.

4.10 Monitoring and Control
What form of monitoring and system control is required?

- From a central TMC?
- On-site (at controller or on-street master)?
- From multiple TMCs?
- From remote locations (not from remote TMC)?
- From maintenance vehicles?
- By an Integrated Corridor Mobility or other external system?

This refers to the location at which an operator will be able to observe and control the system's operation, using a workstation, laptop, smartphone, etc. It does not refer to the physical location at which the system's equipment is located. The alternatives typically required by an agency include: from a workstation at a TMC; via laptop connected directly to a local signal controller or an on-street master; or from another remote location, such as elsewhere within the agency (such as the signal shop or traffic engineer's desk) via a WAN, or via the internet.

Describe this need in broad terms. The details will be examined as the requirements are defined. Sample statements that may be used in this chapter are contained in the Concept of Operations samples table (Appendix B).

4.11 Performance Reporting
Your existing system may have the capability to provide numerous reports about its performance and day-to-day operation. These reports support management of the system, assessment of its performance, trouble shooting, and legal records. An adaptive system may replace the existing system, be a stand-

alone addition or be integrated with your existing system; this may affect the reports and records that are available to you.

Do you want...

- To report measures of performance against the system's objective functions?
- To report measures of performance against your agency's mobility objectives?
- Real time logging of system status?
- Recording of events and data generated by the system (if so, what retention policies)?
- Data storage and data analysis, within or external to the system?

Performance Measurement
There are two parts to performance measurement:

- Is the system optimizing performance in terms of its objective function?
- Is the effect of the system on traffic flows meeting your mobility objectives?

System Operation
The adaptive system will make calculations, and take actions based on the results of those calculations. To what extent do you wish to see the results of those calculations, which may be useful in calibration and fine tuning? For example, a system that calculates a requested cycle length for each intersection in a group, based on measured occupancy may report the occupancy of each loop, the requested cycle length at each intersection, the selected cycle length for the whole group, and the constraints that were applied in selecting that cycle length.

A system that adjusts phase times based on whether a phase gaps out or runs to its maximum may report the actual green time, the permitted maximum green time, the next maximum green time, and any constraint that influenced the calculation of the next maximum green time.

Mobility Objectives
How are your agency-wide mobility objectives defined? If mobility is defined in terms of travel times, level of service and throughput, do you expect the system to report traffic conditions in those terms? If the objective function is to minimize delay or stops, are delay or stops reported by the system?

Other parameters that may be useful in reporting performance include: arrivals on green and red; green time utilization; and measured and estimated queue lengths. While these are useful, you should carefully describe your need in terms of the traffic or mobility performance you must report. However, you should also define the need in terms of the signal timing objectives you set for the system, such as minimizing phase failures, minimizing stops along a specified route, or controlling maximum delay to different users.

Real-Time Logging
Real time logging involves recording such items as phase times, plan in effect, transitions, detections, cycle length, offsets and all alarms and events at the time they occur. It may also log the results of calculations and parameters that control the adaptive systems operation. In this way it is possible to see in real time, on a second by second or cycle by cycle basis the status of each signal. To what extent do you need this data available as it occurs or is collected, at the end of the current cycle (if applicable) or within a specified time interval?

Data Storage
What operational data do you need recorded and stored, and for what period of time must it be available within the system? Do you have an external data storage and analysis facility with which the adaptive system must interface and transfer data? What data is required by that system, and at what frequency must this data be transferred?

External Interfaces
Will the performance measures be passed to another system, such as an Integrated Corridor Management decision support system, or a real-time regional traffic conditions map? Describe this need in broad terms. The details will be examined as the requirements are defined. Sample statements that may be used in this chapter are contained in the Concept of Operations samples table (Appendix B).

An important guide for this issue is the regional ITS architecture. Show how any external interfaces are consistent with the regional ITS architecture, and identify any changes to that architecture that will be required a result of stakeholder agreements made during the preparation of this Concept of Operation.

Use Historical Data To Recreate Events
Use historical logged data to study and investigate past situations. This would be used to address complaints such as: "I sat at this approach for 30 minutes waiting for a green light" or "I had to stop at a red light at every intersection between here and there." This information is also important to determining if the system is responding to situations and input data as according to design expectations. Data warehousing is typically included in the regional ITS architecture. In this section, describe how the adaptive system will need to integrate with any relevant data warehouse.

4.12 Failure Notification

Your existing system has the capability to detect failures and report them in real time to appropriate staff for attention. Some adaptive systems require the operator to be logged in to the system to retrieve alarms and alerts, while others push this information out in real time.

How will notification of failure of the adaptive system be managed?

- Report the alarms and alerts directly to the operations and maintenance staff
- Interface with another system, which will in turn report alarms and alerts through its notification system

Report Directly
In this arrangement, the adaptive system would have its own system to notify maintenance and operations staff of alarms and alerts that it generates.

Report to Interfaced System
In this arrangement, the adaptive system would send its alarms and alerts to another system that has a procedure for notification of maintenance and operations staff. This may be a parallel traffic management system, or a separate maintenance management system.

Describe this need in broad terms. The details will be examined as the requirements are defined. Sample statements that may be used in this chapter are contained in the Concept of Operations samples table (Appendix B).

4.13 Preemption and Priority

In this section, you define the types of preemption and lower level priority that exist and need to be maintained, or will be required in the future.

Is preemption or transit priority required to operate during adaptive operation?

What forms of signal preemption and transit priority will be required?

- Railroad preemption
- Emergency vehicle preemption
- Light Rail Transit priority
- Bus priority

Railroad Preemption

How will railroad preemption be accommodated by the adaptive system? Does the adaptive system acknowledge the preemption and have a suitable recovery process? Describe the preemption operation you need during adaptive operation. Include any special rules you use, such as which phase to serve during and after the preemption, and any additional logic based on measured queuing or other traffic conditions.

Emergency Vehicle Preemption

Does the adaptive system acknowledge and permit emergency vehicle preemption? Does it have a suitable recovery process? Describe the EV preemption operation you need during adaptive operation. Include any special rules you use, such as which phase to serve during and after the preemption, and any additional logic based on measured queuing or other traffic conditions.

Transit Priority

There are several different methods of providing transit priority for buses and light rail. Some are entirely locally based, with communication directly between the transit vehicle and the signal controller, and all priority logic is resident in the local controller.

Other methods of implementing bus priority involve a separate system that determines the response to a priority request and communicates with the local controller. Will the adaptive system be required to communicate with an external priority system in order to provide transit priority? Describe the transit priority operation you need during adaptive operation.

Many LRT systems include logic and hardware to coordinate the interaction between the LRT vehicles and the traffic signal indications. Many of these systems also have special dedicated LRT phases within the traffic signal sequences. Some actually drive the traffic signal operations. Describe the transit priority operation you need during adaptive operation. Include any special rules you use, such as which phase to serve during and after the priority, and any additional logic based on measured queues or other traffic conditions.

Preemption / Priority Frequency

Document the number of preemption/priority calls that will occur under varying circumstances. Document any specific implementation rules that will need to be accommodated by the adaptive system.

Describe this need in broad terms. The details will be examined as the requirements are defined. Sample statements that may be used in this chapter are contained in the Concept of Operations samples table (Appendix B).

4.14 Failure and Fallback Modes

This section will describe the behavior of the adaptive system and the non-adaptive components of the system in the event of failures of system elements that are important to efficient and reliable adaptive operation. Answer the following question.

What failure modes are required?

In the event that the adaptive operation cannot continue, should the operation revert to an existing TOD coordinated operation or free operation of local signals? If TOD coordination is required, should this be under control of a central system, an on-street master or local intersection controller database? Mention the extent to which detector failures need to be accommodated based on your agency's ability to maintain detectors in an operable state.

Describe this need in broad terms. The details will be examined as the requirements are defined. Sample statements that may be used in this chapter are contained in the Concept of Operations samples table (Appendix B).

4.16 Definition and Application of Constraints

Now is the time to identify constraints that may limit the choices that are open to you. These constraints will either be accepted or you can choose to overcome the constraint. This will involve determining what the financial, resource or political cost to overcome the constraint will be. You will then be in a position to evaluate the trade-off between the benefits that will not be realized if the constraint is accepted and the cost of removing the constraint to achieve those benefits.

The following sections describe constraints that may be applicable to your situation.

Infrastructure

Your organization may have existing policies in place that affect the equipment that may be purchased. There may be constraints placed on your organization by funding agencies. These should be identified, the implications for system selection quantified and the consequences of varying from those policies documented. Examples of infrastructure policies that may be applicable include:

- Controller type
- Detector type
- Signal system
- Communications media and protocols
- Cabinet type/space

Management and Human Resources

There are often explicit and implicit management and human resource policies that may become constraints. Great care should be taken with this category of constraint. There is often resistance to change that stems from lack of understanding and the "fear of the unknown." There are also many misconceptions about the nature of adaptive systems upon which staff at all levels form opinions that are difficult to change. Examples of management and HR policies that may be applicable include:

- Staff fears that their jobs will be reduced or made obsolete by the adaptive system
- Staff fears of additional workload without reward
- Agency hiring policies
- Agency training policies

Financial Constraints

Your available financial resources may have an impact on the needs that you are able to accommodate within your requirements. If you are preparing this Concept of Operations at the planning stage, in order to develop and program a project, then your financial constraint may be the maximum funding you expect is possible. This may be spread over several program years, which will give you guidance on when you could schedule the project, or it may give you guidance on how the project could be implemented in phases.

If you are preparing this Concept of Operations after a project has been programmed and budgeted, then this will give you guidance on the how complex the system could be, the geographical area that may be covered by the project, the extent to which the support environment can be implemented. If capital funds are limited, careful consideration should be given to the impact these decisions (limiting work to limit initial capital cost) will have on both short and long term operations and maintenance costs and efficiency.

Complexity

With what level of complexity are you comfortable? You will need to consider the level of complexity of the various adaptive systems and relate them to the ability of your organization to handle that complexity. While it is difficult to quantify this factor, a useful benchmark for comparison that will be understood by most organizations that are considering adaptive operation is the level of complexity of conventional time of day coordination systems. Is the level of complexity your organization will be comfortable with:

- Similar to conventional TOD, without traffic responsive pattern selection
- Similar to conventional TOD, with traffic responsive pattern selection
- Somewhat more complex than conventional TOD
- Significantly more complex than conventional TOD

While this does not lead directly to system requirements, it does provide a basis for determining whether the skills and expertise of your existing operations and maintenance staff will be adequate for the selected system. This will lead to further decisions about appropriate qualifications of staff and training requirements.

People

This section will help you quantify the options available for operation and maintenance of the adaptive system within your organization.

- What are the capabilities of your staff (e.g., hardware maintenance in the field (including detection and communication), and in the office; signal timing skills)?

- What is your operations and maintenance structure (e.g., are signal technicians specialists or do they have other electrician duties; are signal maintenance staff under the direction of traffic engineering staff, or separate public works organization; are maintenance staff in-house or contracted)?
- How is maintenance funded? Does it have a self-sustaining structure or does some maintenance need to be built into purchase price and warranty?

While this does not lead directly to system requirements, it does provide a basis for determining whether other organizational, administrative and financial changes will be required to accommodate the adaptive operation.

Hardware and Software Constraints

This section will help you understand the tradeoffs between flexibility to choose from all available products and limiting that flexibility to use a particular hardware or software product along with the adaptive system. In some cases, state agencies require the acquisition of traffic signal controllers based on a statewide selection. If you are a local agency and are subject to this requirement, then this constraint may be unavoidable.

If the requirement imposed by a state agency is in the form of an incentive, however, then the value of the incentive should be weighed against the consequences as revealed by this constraint preventing you from answering the questions in this document as you would wish based on your operational objectives.

Some agencies, however, might consider the acquisition of products that work with their favorite signal control equipment, or their current equipment, to avoid having to replace it or to avoid the complexity of using different signal controllers. If this applies to you, then consider whether this constraint is preventing you from answering the questions in this document as you would wish based on your operational objectives. If the constraint can be relieved, you may enjoy the benefits of a wider more competitive selection and of selecting requirements that are more closely linked to your objectives.

Schedule

While you may have made commitments to others about the date by which your system will be operational, this should not be imposed as a hard constraint that could potentially compromise your selection of the most appropriate system. However, examples of schedule constraints that would be appropriate to impose on your procurement are:

- A major event for which the proposed system is required to efficiently control traffic
- A hard date by which funding would be withdrawn if sufficient progress is not demonstrated

Although it is appropriate to document these dates in the Concept of Operation, they should not be directly reflected in the System Requirements.

Chapter 5: Envisioned Adaptive System Overview

This chapter is an overview of the envisioned adaptive system. It is a high-level description that will describe the main features and capabilities, other systems with which it will be interfaced, and the scope of its coverage. You should describe its conceptual architecture at a block diagram level with a high-level data flow diagram. This should not show design details.

This description should reflect the needs that are described in the previous chapter. It should illustrate, either graphically or in words, each of the following categories of needs that are relevant:

- Network characteristics
- Type of adaptive operation
- Interfaces to other systems

A good way of illustrating the system is to draw out the activities undertaken by stakeholders in a particular situation, and highlight those that are anticipated to be automated with the adaptive operation. An example of such a diagram is illustrated in Figure 9.

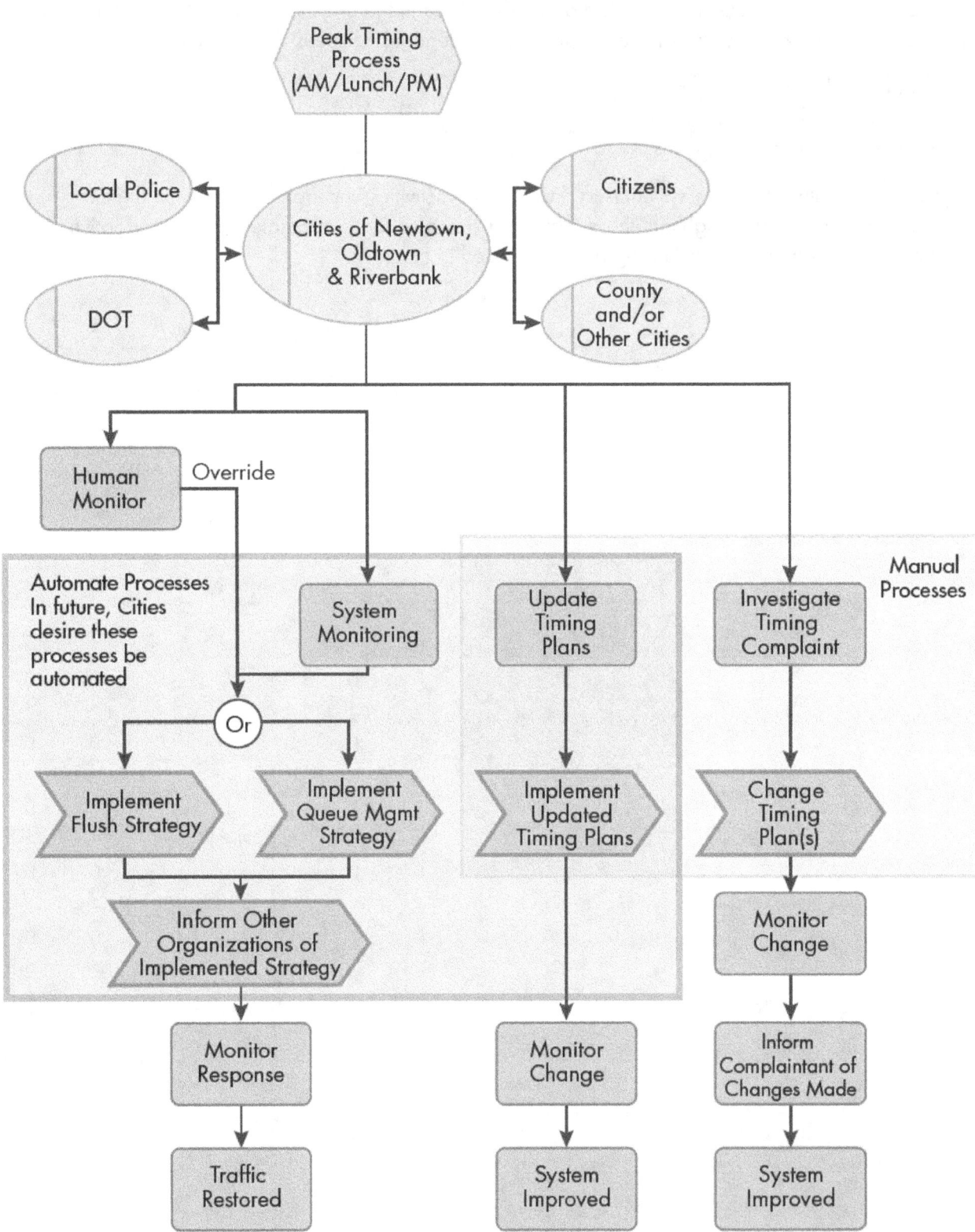

Figure 9. Sample System Block Diagram

Chapter 6: Adaptive Operational Environment

This chapter describes both the operational environment and the physical environment within which the adaptive system will operate.

6.1 Operational Environment

Describe the stakeholders. These should include all existing stakeholders who have an influence on the operation of the existing signal system. This will include traffic engineers involved in signal timing, TMC operations staff, and staff of other agencies whose operation and duties may be affected by the envisioned adaptive system.

The activities related to adaptive operation should be described, such as preparation of timing parameters, implementation and fine tuning, system performance monitoring, and inter-agency staff interactions.

The organizational structure should be described, highlighting any changes from the existing arrangements that are envisioned. An overview of the qualifications and experience of personnel should be presented along with clear definition of any roles and responsibilities that would be undertaken by contractors, vendors, consultants and staff of other agencies.

Sample statements that may be used in this section are contained in the Concept of Operations samples table (Appendix B).

6.2 Physical Environment

You should describe the facilities within which equipment and personnel will be housed, additional furniture and equipment that will be required, new computing hardware and software that will be required, operational procedures for operating the system and any additional support that will be need.

For example, describe whether the equipment will be located in a TMC, at City Hall, at the corporation yard or signal shop and/or in the field. Will field equipment need to be field hardened or located within an air-conditioned environment? Will existing power supplies be adequate or will additional service, UPS and battery backups be required?

Will the operators be on duty or available 24/7 or during limited hours? Describe their required experience, skills and additional training needs.

Sample statements that may be used in this section are contained in the Concept of Operations samples table (Appendix B).

Chapter 7: Adaptive Support Environment

This chapter describes the current and planned physical support environment. Describe what support equipment, personnel, training and procedures currently existing, and explain those that need to be acquired or implemented.

Describe any additional test equipment and repair tools that will be needed to support the adaptive operation. Where will test equipment be located? Will system simulators be required (e.g., hardware and software to allow simulation of traffic and signal timing before new timing data goes live). Will a development server be required to set up controller firmware before deployment, and to test system upgrades and modifications before deployment?

Describe additional staff or contractors who will not be involved in the day-to-day operation of the system, but will be needed to support the operators and maintenance staff. This should include staff from the system vendor and/or consultants, who will provide additional on-going training, periodically audit the system setup and performance and support expansion of the system in the future.

Where multiple agencies are involved, describe the support that will be provided by or to other agencies. This should include any existing or proposed memoranda of understanding or operations and maintenance agreements that will affect the adaptive system, or will need to be modified to include reference to the adaptive system. This may include modifying the policies and procedures of those agencies in addition to developing new policies and procedures within your agency.

During the life of the adaptive system, will any equipment or supplies specifically related to the adaptive system need to be disposed of? Can this be done using existing procedures and protocols, or will additional arrangements be needed? At the end of useful life of the system, will any special disposal arrangements be required?

Sample statements that may be used in this chapter are contained in the Concept of Operations samples table.

Chapter 8: Proposed Operational Scenarios Using an Adaptive System

The purpose of this chapter of the Concept of Operations document is to provide examples that illustrate how the system will be expected to operate and interface with the operators in typical circumstances. It is not intended to comprehensively describe the operation under all conditions. **It is intended to illustrate to vendors, managers and decision-makers alike how you see your objectives being met by the system.** This description is practically oriented and takes into account the practical limitations of available systems, which you expect to be live with. It should not be a description of how you would like some imagined system to operate with no regard for the practical limitation of candidate systems.

Each statement in a scenario should relate to a user need, although not all needs will be further described in a scenario. The statements in the description of each scenario do not directly generate requirements. Requirements are only generated by needs. The scenarios simply provide examples of how the system meets some of the needs.

Once you have written the scenarios, if you are not satisfied that they describe an operation that will be adequate, you should then review your needs statements. If you wish to describe elements of the proposed operation that are not described by needs, then additional needs should be enunciated.

8.1   How to Construct a Scenario

Each scenario should describe a unique set of circumstances, applying to one type of location, one set of traffic conditions with one set of appropriate activities by stakeholders, and one response by the ASCT system. A scenario should include statements about each of the following elements:

- Road network on which the scenario occurs
- Traffic conditions that must be accommodated
- The operational objectives that should be satisfied in this situation
- The coordination and signal timing strategies that should be applied by the ASCT system to satisfy those objectives.

If all the scenarios relate to the same section of road network, the description does not need to be repeated within each scenario. If the network is the same as described in the existing conditions, then the description may be deleted from the scenarios.

Example statements are included in the appendix. Each one is linked to an appropriate need. If you edit these statements or create your own, you must ensure that each new statement is matched to at least one of your statements of need, otherwise, you will not have a requirement that will ensure the system allows you to operate as you describe in your scenario.

Example scenarios used in a real project are also provided in Appendix C.

E. SYSTEM REQUIREMENTS GUIDANCE

The chapters required for the System Requirements document are:

1. Scope of System or Sub-system
2. References
3. Requirements
4. Verification Methods
5. Supporting Documentation
6. Traceability Matrix
7. Glossary

This document sets the technical scope of the system to be built. It is the basis for verifying (via the Verification Plan) the system and sub-systems when delivered.

This document must be tailored to your project. All signal system projects need a set of requirements defining what is needed. You will need to decide how extensively to document these requirements. One convenient way to gauge how many requirements to write and/or how much detail to have in the requirements document is to start at the finish line. The following should be asked when starting at the top level of the system:

- What are all the functions needed in order to demonstrate to the agency that the system is doing what it is expected to do?
- How well does the system need to perform the required functions?
- Under what conditions does the system need to operate?

Each of these tests will need a set of requirements. This is done for the system and the sub-systems. For simple systems one or two pages of requirements may be sufficient to fully define what the system is to do. In more complex systems this could be 10 to 20 pages or more.

Another factor that drives the number of requirements and depth of detail that needs to be written is the extent to which commercially available products are used. These products have their own specifications. For the non-adaptive requirements, it may be sufficient to reference the existing product specifications after they have been carefully reviewed to determine if the product will meet the agency's intended need. For example, the traffic control systems that are on the market have sufficient documentation to cover the majority of the non-adaptive functions that are required. The additional requirements would be for any modifications or enhancements needed. However, great care must be taken when referencing existing commercial product specifications to ensure the wording does not unnecessarily or unintentionally limit compliance to a single system when more than one is capable of providing the required functionality.

When choosing from available products, your requirements don't need to be as detailed as they would be when developing a new system. These model documents apply only to the former situation. If your needs lead you to decide that new software must be developed, the project will be of sufficient scope and risk to warrant a more detailed and customized system engineering process than is provided by these model documents.

Once the requirements document has been completed, use this checklist to confirm that all critical information has been included.

- ✓ Is there a definition of all the major system functions?
- ✓ With each function of the system, is there a set of requirements that describes: what the function does, and under what conditions (e.g., environmental, reliability, and availability.)
- ✓ Are all terms, definitions, and acronyms defined?
- ✓ Are all supporting documents such as standards, concept of operations, and others referenced?
- ✓ Does each requirement have a link (traceability) to a higher level requirement of a user-specified need or scenario?
- ✓ Is each requirement concise, verifiable, clear, feasible, necessary, unambiguous, and technology (vendor) independent?
- ✓ Are all technology dependent requirements identified as constraints?
- ✓ Does each requirement have a method of verification defined?
- ✓ Does each requirement trace to a verification case?

1 Scope of System or Sub-system (Chapter 1 of the System Requirements)

This chapter is a brief overview of the system and statement of the purpose of this document. Briefly describe the contents, intention and audience for this document. Summarize the history of system development, the proposed operation, and maintenance. Identify the project stakeholders, acquiring agency, users and support agencies. Identify current and planned operating sites.

2 References (Chapter 2 of the System Requirements)

This chapter identifies all standards, policies, laws, Concept of Operations document, concept exploration documents and other reference material that are needed to support the requirements.

3 Requirements (Chapter 3 of the System Requirements)

This chapter lists all the requirements necessary to define the proposed adaptive system. Each requirement should be clear and concise, verifiable, feasible, necessary, unambiguous and technology independent. Each requirement should have a single statement. DO NOT use terms such as "and", "but", "except" and other modifiers that combine more than one thought into a single requirement. The requirements listed in the sample table are organized under the following headings:

- Network characteristics
- Type of operation
- External/Internal interfaces
- Crossing arterials and boundaries
- Access and security
- Data log
- Advanced controller operation
- Pedestrians

- Special functions
- Detection
- Railroad and EV preemption
- Transit priority
- Failure events and fallback
- Software
- Training
- Maintenance, support and warranty
- Schedule
- Performance measurement, monitoring and reporting.

In general, each of the sample requirements falls into one of the following categories, although they are not expected to be organized in this manner:

- *Functional requirements* (What the system shall do)
- *Performance requirements* (How well the requirements should perform)
- *Non-Functional requirements* (Reliability, safety, environmental)
- *Enabling requirements* (Production, development, testing, training, support, deployment, and disposal). This can be done through references to other documents or can be explicitly defined in these requirements.
- *Constraints* (e.g., Technology, design, tools, and/or standards)

If this document is describing a sub-system, the following may also be required:
- *Interface requirements* (Definition of the interfaces)
- *Data requirements* (Data elements and definitions)

Sample requirements that may be used in the Requirements document are included in the System Requirements samples table. Each of these is directly related to one or more statements of need in the Concept of Operation.

4 Verification Methods (Chapter 4 of the System Requirements)

In this chapter, identify one of the following methods of verification for each requirement.

- *Demonstration* is used for a requirement that the system can demonstrate without external test equipment.
- *Test* is used for a requirement that requires some external piece of test equipment (such as logic analyzer and volt meter.
- *Analyze* is used for a requirement that is met indirectly through a logical conclusion or mathematical analysis of a result. For example, algorithms for congestion: the designer may need to show that the requirement is met through the analysis of count and occupancy calculations in software or firmware.
- *Inspection* is used for verification through a visual comparison. For example, quality of welding may be done through a visual comparison against an in-house standard.

Do not describe how, when or where the verification will be performed. This is separately covered in the Verification Plan, which is discussed in a later section.

5 Supporting Documentation (Chapter 5 of the System Requirements)

This optional chapter is a catch-all for anything that may add to the understanding of the Requirements and cannot be logically located elsewhere. Examples of supporting documents include: diagrams, analysis, memos, stakeholders contact list and published documents related to similar projects.

6 Traceability Matrix (Chapter 6 of the System Requirements)

This is a table that traces the requirements in this document to the needs expressed in the Concept of Operation. Table 2 contains an extract from a traceability matrix used in a project with multi-level requirements (Business, User and Functional).

Table 2. Example Traceability Matrix

User Requirement	User Requirement Description	Satisfies Business Requirement	Satisfied By Functional Requirement
TO	Traffic Operations		
TO-1	System operator needs to support vehicle, pedestrian and transit traffic mobility	1, 4, 8	3, 4, 9, 10
TO-2	System operator needs to identify changing traffic conditions	2, 5	13, 14, 15, 16
TO-3	System operator needs to adjust the operation based on the traffic conditions	2, 5	1, 2, 16, 20
TO-4	System operator needs to reduce the delay and travel time of the priority movements of each intersections	2	1, 5, 6, 7, 8
TO-5	System operator needs to minimize adverse effects caused by preemption and unexpected events	2	1, 11, 12, 19
H	Hardware		
H-6	The agency needs to reuse existing equipment where possible without reducing the service life of the system	10	17, 18

F. VERIFICATION PLAN GUIDANCE

The chapters required for the Verification Plan are:

1. Purpose of Document
2. Scope of Project
3. Referenced Documents
4. Conducting Verification
5. Verification Identification

This verification plan describes the activity of verifying that the system being built satisfies all the requirements set out in the requirements documents. The verification documents will include:

- A plan to initially lay out the specific verification effort
- The verification plan that defines the detailed mapping of the requirements to verification cases
- A report on the results of the Verification activities

To ensure that all requirements are verified by this activity, trace each requirement into a verification case, then trace this in turn into a step in the Verification procedure.

The Verification Plan does not need to include verification procedures. These may be prepared by the vendor, but must be reviewed by the agency to ensure each verification case will be tested and appropriate results recorded. In relatively simple cases, both the Verification Plan and the procedures may be prepared by the vendor. In this situation the agency must ensure that each requirement is mapped to verification test.

Preparation of a stand-alone verification plan is strongly advised if:

- The system is complex
- There are several separate verification activities being performed on the system
- Multiple deployment sites are involved
- Multiple stakeholders have to be satisfied

There is also the question of how comprehensive to make the verification effort. It is impossible to validate all possible combinations of actions under all possible operational situations. A good rule of thumb is: if it was important enough to write down as a requirement, then it should be verified, at least once. In-process[4] verification performed on the needs and requirements will help ensure that the correct requirements are being verified.

Once the verification plan is completed, use the following checklist to ensure all critical information has been included.

- ✓ Does the Verification Plan answer all the questions of who, what, where, and when?
- ✓ Does the Verification Plan make clear what needs to happen if a failure is encountered?
- ✓ Does the Verification Plan document the configuration of the hardware, software?
- ✓ Are all requirements traced to a verification case?

---

[4] In-process verification is reviewing the needs and requirements during the definition stage by the stakeholders to ensure that all the needs have been identified and traced to appropriate requirements and have been reviewed for completeness for each of the needs.

1 Purpose of Document (Chapter 1 of the Verification Plan)

This chapter identifies the type of verification activity to be performed within this Verification Plan. For instance, this activity may validate the entire system, a sub-system, the deployment at a site, a burn-in, or any other verification activity called for in a relevant Program Plan.

2 Scope of Project (Chapter 2 of the Verification Plan)

This chapter gives a brief description of the planned project and the purpose of the system to be built. It also describes the environment in which the project operates. It identifies the organization structures that encompass all stakeholders. It also gives a brief description of the role to be played by each stakeholder. This includes ad hoc and existing management work groups and multi-disciplinary technical teams that should be formed to support the project.

3 Referenced Documents (Chapter 3 of the Verification Plan)

This is a list of all documents used in the preparation of this Verification Plan. This almost always includes the Project Plan (if one was written), and the applicable Requirements Documents. However, reference of other documents, such as descriptions of external systems, standards, a Concept of Operations, and manuals may also need to be included.

4 Conducting Verification (Chapter 4 of the Verification Plan)

This chapter provides details on how verification is accomplished. It defines: who does the verification; when and where it is to be done; the responsibilities of each participant before, during and after verification; the deployed hardware and software configuration; and the documents to be prepared as a record of the verification activity. It is also important to define how anomalies are to be handled (that is, what to do when a failure occurs during verification).

In general, the following information should be included in this section:

- A description of the participating organizations and personnel and identification of their roles and responsibilities. This may include for example, a verification conductor, verification recorder, operators, and/or engineering support.
- Identification of the location of the verification effort, that is, the place, or places, where the verification must be observed.
- The deployed hardware and software configuration for all of the verification cases, including hardware and software under verification and any supporting equipment, software, or external systems. Several configurations may be necessary.
- Identification of the documents to be prepared to support the verification, including Verification Procedures, a Verification Report and descriptions of special equipment and software.
- Details on the actual conduct of verification, including:
  - Notification of participants
  - Emphasis on the management role of the verification conductor
  - Procedures for approving last minute changes to the procedures

- The processes for handling a failure, including recording of critical information, determination of whether to stop the verification, restart, or skip a procedure, resolution of the cause of a failure (e.g. fix the software, reset the system, and/or change the requirements), and determination of the re-verification activities necessary as a result of the failure.

5   Verification Identification (Chapter 5 of the Verification Plan)

This section identifies the specific verification cases to be performed. A verification case is a logical grouping of functions and performance criteria (all from the Requirements Documents) that are to be verified together. For instance, a specific verification case may cover all the control capabilities to be provided for control of the adaptive control system. There may be several individual requirements that define this capability, and they all are verified in one case. The actual grouping of requirements into a case is arbitrary. They should be related and easily combined into a reasonable set of procedure actions.

Each case should contain at least the following information:

- A description name and a reference number.
- A description of the objective of the verification case, usually taken from the wording of the requirements, to aid the reader understanding the scope of the case.
- A complete list of requirements to be verified or traceability to the requirements in the requirements document. Since each requirement has a unique number, they can be accurately and conveniently referenced without repetition.
- Any data to be recorded or noted during verification, such as expected results of a step. Other data, such as a recording of a digital message sent to an external system, may be required to validate the performance of the system.
- A statement of the pass/fail criteria. Often, this is just a statement that the system operates per the need or scenario.
- A description of the verification configuration. That is a list of the hardware and software items needed for verification and how they should be connected (in most cases this is the deployed system configuration). Often, the same configuration is used for several verification cases.
- A list of any other important assumptions and constraints necessary for conduct of the verification case.

G. VALIDATION PLAN GUIDANCE

The chapters required for the Validation Plan document are:

1. Purpose of Document
2. Scope of Project
3. Referenced Documents
4. Conducting Validation
5. Validation Identification

This document describes the activity of validation that the system being built meets the user needs and scenarios developed in the concept of operations. The validation documents will generally include three levels of validation documents:

- A plan to initially lay out the specific validation effort
- The user's/operator's manual and/or a validation plan that defines the detailed operational procedures
- A report on the results of the validation activities

To ensure user needs and scenarios are validated by this activity, trace each need and scenario into a validation case, then into appropriate steps in the validation procedure.

A separate Validation Plan and procedures may be minimal for the simplest projects, especially where the system is commercially available and does not involve any custom software development, and where the agency staff have a very clear understanding of the purpose of the system. Preparation a validation plan is strongly advised if:

- The system is more complex
- There are several separate validation activities
- Multiple deployment sites are involved
- Multiple stakeholders have to be satisfied

There is also the question of how comprehensive to make the validation effort. It is impossible to validate all possible combinations of actions under all possible operational situations. A good rule of thumb is: if it was important enough to write down as a need or scenario, then it should be validated, at least once. This may not, for example, validate all possible failure mode conditions or all possible incident scenarios. In-process[5] validation performed on the needs will help ensure that end to end validation of the system will meet the stakeholder needs.

Once the Validation Plan has been prepared, use this checklist to ensure all critical information has been included.

- ✓ Does the Validation Plan answer all the questions of who, what, where, and when?
- ✓ Does the Validation Plan make clear what needs to happen if an unexpected situation or a failure is encountered?

---

[5] In-process validation is reviewing the needs and requirements during the definition stage by the stakeholders to ensure that all the needs have been identified and traced to appropriate requirements and have been reviewed for completeness for each of the needs.

- ✓ Does the Validation Plan document the configuration of the hardware and software?
- ✓ Are all applicable needs and scenarios traced to a validation case?

1 Purpose of Document (Chapter 1 of the Validation Plan)

This chapter identifies the type of validation activity to be performed within this Validation Plan. For instance, this activity may validate the entire system, a sub-system, the deployment at a site, a burn-in, or any other validation activity called for in a relevant Program Plan or SEMP.

2 Scope of Project (Chapter 2 of the Validation Plan)

This chapter gives a brief description of the planned project and the purpose of the system to be built. Special emphasis is placed on the project's user needs and issues that must be addressed and validated.

This chapter also describes the environment in which the project operates. It identifies the organization structures that encompass all stakeholders. It also gives a brief description of the role to be played by each stakeholder. This includes ad hoc and existing management work groups and multi-disciplinary technical teams that should be formed to support the project.

3 Referenced Documents (Chapter 3 of the Validation Plan)

This is a list of all documents used in the preparation of this Validation Plan. This almost always includes the Project Plan, the SEMP (if one was written), and the applicable Requirements Documents. However, reference of other documents, such as descriptions of external systems, standards, a Concept of Operations, and manuals may need to be included.

4 Conducting Validation (Chapter 4 of the Validation Plan)

This chapter provides details on how validation is accomplished. It defines: who does the validation; when and where it is to be done; the responsibilities of each participant before, during, and after validation; the deployed hardware and software configuration; and the documents to be prepared as a record of the validation activity. This chapter defines how anomalies are to be handled (that is, what to do when an unexpected situation or a failure occurs during validation).

In general, the following information should be included in this chapter:

- A description of the participating organizations and personnel and identification of their roles and responsibilities. This may include for example, a validation conductor, validation recorder, operators, and/or engineering support.
- Identification of the location of the validation effort, that is, the place, or places, where the validation must be observed.
- The deployed hardware and software configuration for all of the validation cases, including hardware and software under validation and any supporting equipment, software, or external systems. Several configurations may be necessary.

- Identification of the documents to be prepared to support the validation, including Validation Procedures, a Validation Report and descriptions of special equipment and software.
- Details on the actual conduct of Validation, including:
  - Notification of participants
  - Emphasis on the management role of the validation conductor
  - Procedures for approving last minute changes to the procedures

The processes for handling a failure, including recording of critical information, determination of whether to stop the validation, restart, or skip a procedure, resolution of the cause of a failure (e.g. fix the software, reset the system, and/or change the requirements), and determination of the re-validation activities necessary as a result of the failure.

5   Validation Identification (Chapter 5 of the Validation Plan)

This chapter identifies the specific validation cases to be performed. A validation case is a logical grouping of functions and performance criteria (all from the Concept of Operations Documents) that are to be validated together. For instance, a specific validation case may cover all the control of traffic during the AM peak hour. There may be several individual objectives and associated performance measures that define this capability, and they all are validated in one case. The actual grouping of objectives into a case is arbitrary. They should be related and easily combined into a reasonable set of procedure actions.

Each case should contain at least the following information:

- A description name and a reference number.
- A complete list of needs and scenarios to be validated. For ease of tracing of needs and scenarios into the Validation Plan and other documents, the needs and scenarios are given numbers. They can be accurately and conveniently referenced without repetition.
- A description of the objective of the validation case, usually taken from the wording of the need or scenario, to aid the reader understanding the scope of the case.
- Any data to be recorded or noted during Validation, such as expected results of a step. Other data, such as a recording of a digital message sent to an external system, may be required to validate the performance of the system.
- A statement of the pass/fail criteria. Often, this is just a statement that the system operates per the need or scenario.
- A description of the validation configuration. That is a list of the hardware and software items needed for validation and how they should be connected (this should be in most cases the deployed system configuration). Often, the same configuration is used for several validation cases.
- A list of any other important assumptions and constraints necessary for conduct of the validation case.

Appendix A
Document Templates

CONCEPT OF OPERATIONS

PURPOSE OF THIS DOCUMENT

The Concept of Operations is a description of how the system will be used. It is non-technical, and presented from the viewpoints of the various stakeholders. This provides a bridge between the often vague needs that motivated the project to begin with and the specific technical requirements. There are several reasons for developing a Concept of Operations.

- Get stakeholder agreement identifying how the system is to be operated, who is responsible for what, and what the lines of communication are
- Define the high-level system concept and justify that it is superior to the other alternatives
- Define the environment in which the system will operate
- Derive the user needs that should be accommodated by the proposed system
- Provide scenarios that describe how the system is expected to operate in practical situations
- Provide criteria to be used for validation of the completed system

CHECKLIST: CRITICAL INFORMATION

- ✓ Is the reason for developing the system clearly stated?
- ✓ Are all the stakeholders identified and their anticipated roles described? This should include anyone who will operate, maintain, build, manage, use, or otherwise be affected by the system.
- ✓ Are alternative operational approaches (such as traffic responsive or time of day coordination) described and the selected approach justified?
- ✓ Is the external environment described? Does it include required interfaces to existing systems?
- ✓ Is the support environment described? Does it include maintenance?
- ✓ Is the operational environment described?
- ✓ Are there clear and complete descriptions of normal operational scenarios?
- ✓ Are there clear and complete descriptions of maintenance and failure scenarios?
- ✓ Do the scenarios include the viewpoints of all involved stakeholders? Do they make it clear who is doing what?
- ✓ Are all constraints on the system identified?

CONCEPT OF OPERATIONS TEMPLATE

Section	Contents
Title Page	The title page should follow the Agency's procedures or style guide. At a minimum, it should contain the following information: CONCEPT OF OPERATIONS FOR THE (insert name of project) - (insert name of sponsoring agency) Date that the document was formally approved The organization responsible for preparing the document Internal document control number, if available Revision version and date issued
1.0 Scope	This chapter is a brief statement of the purpose and scope of this document; and the purpose and scope of the project. This will briefly describe contents, intention and audience. A few paragraphs will normally suffice. This chapter also gives a brief overview of the system to be built. It includes its purpose and a high-level description. It describes what area will be covered and which agencies will be involved, either directly or through interfaces. A few paragraphs and a map will usually suffice.
2.0 Referenced Documents	This chapter is a place to list any supporting documentation used in preparing this document and other resources that are useful in understanding the operations of the system. This could include any documentation of current operations and any strategic plans that drive the goals of the system under development.
3.0 User-Oriented Operational Description	This chapter is a brief description aimed at non-technical readers who need an understanding of the current system or situation. It should briefly say what the existing system is, how it is currently used, what you are currently able to achieve with the system and (most importantly) what you want to do that can't currently be achieved with the system.
4.0 Operational Needs	This chapter lists all the needs of the stakeholders that are expected to be accommodated by the proposed system. These are expressed in non-technical terms that can be understood by all readers. The needs must have a clear nexus with the goals and objectives. This list of needs is the basis for preparation of the detailed requirements. Each requirement must satisfy at least one need, and each need must be supported by at least one requirement.

Section	Contents
5.0 System Overview	This chapter is an overview of the envisioned adaptive system. It is a high level description that will describe its goals and objectives, the main features and capabilities, other systems with which it will be interfaced, the users of the system, and the scope of its coverage. Describe its conceptual architecture at a block diagram level, with a high-level data flow diagram. This should not show design details.
6.0 Adaptive Operational Environment	This section describes the physical operational environment in terms of facilities, equipment, computing hardware, software, personnel and operational procedures necessary to operate the deployed system. For example, it will describe the personnel in terms of their expected experience, skills and training, typical work hours, and other activities that must be or may be performed concurrently.
7.0 Adaptive Support Environment	This describes the current and planned physical support environment. This includes facilities, utilities, equipment, computing hardware, software, personnel, operational procedures, maintenance, and disposal. It also includes expected support from outside agencies.
8.0 Operational Scenarios	The purpose of this chapter is to provide examples that illustrate how the system will be expected to operate and interface with the operators in typical circumstances. It is not intended to comprehensively describe the operation under all conditions. It is intended to illustrate to vendors, managers and decision-makers alike how you see your objectives being met by the system. This description is practically oriented and takes into account the practical limitations of available systems. It should not be a description of how you would like some imagined system to operate with no regard for the practical limitation of candidate systems. Each statement in a scenario should relate to a user need, although not all needs will be further described in a scenario. The statements in the description of each scenario do not directly generate requirements. Requirements are only generated by needs. The scenarios simply provide examples of how the system meets some of the needs. The scenarios will need to cover all normal conditions, stress conditions, failure events, maintenance, and anomalies and exceptions.
Appendices	This is a place to put a glossary, notes, and backup or background material for any of the sections. For example, it might include analysis results in support of the concept exploration.

SYSTEM REQUIREMENTS

PURPOSE OF THIS DOCUMENT

This requirements document describes what the system is to do (functional requirements), how well it is to perform (performance requirements), under what conditions it will perform (non-functional requirements), and what other actions are required in order for the system to become fully operational (enabling requirements). This document does not define how the system is to be built. It primarily defines requirements that are necessary to satisfy the operational needs identified in the Concept of Operations. Each requirement must satisfy at least one of the needs described in the Concept of Operations.

This document sets the technical scope of the system to be built. It is the basis for verifying (via the Verification Plan) the completeness of the system and sub-systems when delivered. Every requirement must be testable and verifiable.

TAILORING THIS DOCUMENT TO YOUR PROJECT

All signal system projects need a set of requirements defining what is needed. The tailoring is in how extensively to document these requirements. One way to gauge how many requirements to write and/or how much detail to have in the requirements document is to start at the finish line. The following should be asked when starting at the top level of the system:

- What are all the functions needed in order to satisfy the agency that the system is doing what it is expected to do?
- How well does the system need to perform the required functions?
- Under what conditions does the system need to operate?

Each of these questions leads to a set of requirements. This is done for the system and any sub-systems. A simple system may be fully defined using only one or two pages of requirements describing what the system is to do. In more complex systems this could be many more pages.

Another factor that drives the extent to which requirements need to be written is the amount of commercially available products that are used. These products have their own specifications, so it may be sufficient to reference them after they have been reviewed to determine if the product will meet the agency's intended need. For example, the traffic control systems that are on the market have sufficient documentation to cover the majority of functions that are required. The additional requirements would be for any modifications or enhancements needed. However, care should be taken to ensure the referenced specifications do not lock you into one vendor when several are capable of satisfying your needs.

CHECKLIST: CRITICAL INFORMATION

- ✓ Is there a definition of all the major system functions?
- ✓ With each function of the system, is there a set of requirements that describes: what the function does, and under what conditions (e.g., environmental, reliability, and availability.)?
- ✓ Are all terms, definitions, and acronyms defined?
- ✓ Are all supporting documents such as standards and concept of operations referenced?
- ✓ Does each requirement have a link (traceability) to a higher level requirement or a user-specified need?
- ✓ Is each requirement concise, verifiable, clear, feasible, necessary, unambiguous and technology independent?
- ✓ Are all technology dependent requirements identified as constraints?
- ✓ Does each requirement have a method of verification defined?
- ✓ Does each requirement trace to a verification case?

SYSTEM REQUIREMENTS TEMPLATE

Section	Contents
Title Page	The title page should follow the Sponsoring agency procedures or style guide. At a minimum, it should contain the following information:
	SYSTEM REQUIREMENTS for (insert name of project) - (insert name of Sponsoring agency)
	Date that the document was formally approved
	The organization responsible for preparing the document
	Internal document control number, if available
	Revision version and date issued
1.0 Scope of System or Sub-system	Provides a system overview and briefly states the purpose of the system
	Describes the general nature of the system
	Summarizes the history of system development, operation, and maintenance
	Identifies the project stakeholders, acquirer, users and support agencies
	Identifies current and planned operating sites
2.0 References	Identifies all needed standards, policies, laws, concept of operations, concept exploration documents and other reference material that supports the requirements.
3.0 Requirements	List all the requirements with which a vendor must comply. Include requirements covering each of the following types: • *Functional requirements* (What the system shall do) • *Performance requirements* (How well the requirements should perform) • *Non-Functional requirements* (Reliability, safety, environmental (temperature) • *Enabling requirements* (Production, development, testing, training, support, deployment, and disposal.) This can be done through references to other documents or embedded in this requirements • *Constraints* (E.g. Technology, design, tools, and/or standards) If this document is describing a sub-system, the following may also be required: • *Data requirements* (Data elements and definitions of the system) • *Interface requirements* (Definition of the interfaces)

Section	Contents
4.0 Verification Methods	For each requirement, identify one of the following methods of verification: • *Demonstration* is a requirement that the system can demonstrate without external test equipment. • *Test* is a requirement that requires some external piece of test equipment. E.g. logic analyzer, and/or volt meter. • *Analysis* is a requirement that is met indirectly through a logical conclusion or mathematical analysis of a result. E.g., algorithms for congestion: the designer may need to show that the requirement is met through the analysis of count and occupancy calculations in software or firmware. • *Inspection* is verification through a visual comparison. For example, quality of welding may be done through a visual comparison against an in-house standard.
5.0 Supporting Documentation	Catch-all for anything that may add to the understanding of the Requirements without going elsewhere (Reference section) Examples: diagrams, analysis, key notes, memos, rationale, stakeholders contact list
6.0 Traceability Matrix	This is a table that traces the requirements in this document to the needs in the Concept of Operation.
7.0 Glossary	Terms, acronyms, definitions

VERIFICATION PLAN

PURPOSE OF THIS DOCUMENT

This verification plan describes the activity of verifying that the system being built satisfies all the requirements set out in the requirements documents. A critical issue is assuring that all requirements are verified by this activity. This is best done by tracing each requirement into a verification test case, then traced into a step in a verification procedure.

TAILORING THESE DOCUMENTS TO YOUR PROJECT

The Verification Plan and procedures may be minimal for the simplest projects, especially where the system is commercially available and does not involve any custom software development, and where the agency staff have a very clear understanding of the purpose of the system. In some cases, the vendor will provide an acceptance test plan for their features and any custom features requested by the agency. The agency should review the vendor's verification plan to ensure all the features are being verified.

Preparation a Verification Plan is advisable if:

- The system is complex
- There are a number of separate verification activities being performed on the system
- Multiple deployment sites are involved
- Multiple stakeholders have to be satisfied

The Verification Plan does not need to include verification procedures. These may be prepared by the vendor, but must be reviewed by the agency to ensure each verification case will be tested and appropriate results recorded. In relatively simple cases, both the Verification Plan and the procedures may be prepared by the vendor. In this situation the agency must ensure that each requirement is mapped to verification test.

CHECKLIST: CRITICAL INFORMATION

- ✓ Does the Verification Plan answer all the questions of who, what, where, and when?
- ✓ Does the Verification Plan make clear what needs to happen if a failure is encountered?
- ✓ Does the Verification Plan document the configuration of the hardware and software?
- ✓ Are all requirements traced to a verification case?

VERIFICATION PLAN TEMPLATE

Section	Contents
Title Page	The title page should follow the Sponsoring agency's procedures or style guide. At a minimum, it should contain the following information: VERIFICATION PLAN FOR THE (insert name of project) - (insert name of Sponsoring agency) Date that the document was formally approved The organization responsible for preparing the document Internal document control number, if available Revision version and date issued
1.0 Purpose of Document	This section identifies the type of verification activity to be performed within this Verification Plan. For instance, this activity may validate the entire system, a sub-system, the deployment at a site, a burn-in, or any other verification activity called for in the Verification Plan.
2.0 Scope of Project	This section gives a brief description of the planned project and the purpose of the system to be built. This section also describes the environment in which the project operates. It identifies the organization structures that encompass all stakeholders. It also gives a brief description of the role to be played by each stakeholder. This includes ad hoc and existing management work groups and multi-disciplinary technical teams that should be formed to support the project.
3.0 Referenced Documents	This is a list of all documents used in the preparation of this Verification Plan. This almost always includes the Project Plan, (if one was written), and the applicable Requirements Documents. Reference to other documents, such as descriptions of external systems, standards, a Concept of Operations, and manuals may also be included.

Section	Contents
4.0 Conducting Verification	This section provides details on how verification is accomplished. It defines: who does the verification; when and where it is to be done; the responsibilities of each participant before, during, and after verification; the deployed hardware and software configuration; and the documents to be prepared as a record of the verification activity. It is also important to define how anomalies are to be handled (that is, what to do when a failure occurs during verification). In general, the following information should be included in this section: - A description of the participating organizations and personnel and identification of their roles and responsibilities. This may include for example, a verification conductor, verification recorder, operators, and/or engineering support. - Identification of the location of the verification effort, that is, the place, or places, where the verification must be observed. - The deployed hardware and software configuration for all of the verification cases, including hardware and software under verification and any supporting equipment, software, or external systems. Several configurations may be necessary. - Identification of the documents to be prepared to support the verification, including Verification Procedures, a Verification Report and descriptions of special equipment and software. - Details of the actual conduct of verification, including: - Notification of participants - Emphasis on the management role of the verification conductor - Procedures for approving last minute changes to the procedures - The processes for handling a failure, including recording of critical information, determination of whether to stop the verification, restart, or skip a procedure, resolution of the cause of a failure (e.g. fix the software, reset the system, and/or change the requirements), and determination of the re-verification activities necessary as a result of the failure.

Section	Contents
5.0 Verification Identification	This section identifies the specific verification cases to be performed. A verification case is a logical grouping of functions and performance criteria (all from the Requirements Documents) that are to be verified together. For instance, a specific verification case may cover all the control capabilities to be provided for control of the adaptive control system. There may be several individual requirements that define this capability, and they all are verified in one case. The actual grouping of requirements into a case is arbitrary. They should be related and easily combined into a reasonable set of procedure actions. Each case should contain at least the following information: - A description name and a reference number - A description of the objective of the verification case, usually taken from the wording of the requirement, to aid the reader understanding the scope of the case - A complete list of requirements to be verified or traceability to the requirements in the requirements document. Since each requirement has a unique number, they can be accurately and conveniently referenced without repetition. - Any data to be recorded or noted during verification, such as expected results of a step. Other data, such as a recording of a digital message sent to an external system, may be required to validate the performance of the system. - A statement of the pass/fail criteria. Often this is just a statement that the system operates per the need or scenario. - A description of the verification configuration. That is a list of the hardware and software items needed for verification and how they should be connected (in most cases this is the deployed system configuration). Often, the same configuration is used for several verification cases. - A list of any other important assumptions and constraints necessary to conduct the verification case.

VALIDATION PLAN

PURPOSE OF THIS DOCUMENT

This document describes the activity of validation that the system being built meets the user needs and scenarios developed in the concept of operations. Usually, for even moderately complex systems, the following three levels of validation documents are prepared:

- A plan to initially lay out the specific validation effort
- The user's/operator's manual and/or a validation plan that defines the detailed operational procedures
- A report on the results of the validation activities

A critical issue is assuring that user needs and scenarios are validated by this activity. This is best done by first tracing each need and scenario into a validation case, then into a step in the validation procedure.

TAILORING THESE DOCUMENTS TO YOUR PROJECT

A separate Validation Plan and procedures may be minimal for the simplest projects, especially where the system is commercially available and does not involve any custom software development, and where the agency staff have a very clear understanding of the purpose of the system. Preparation a validation plan is strongly advised if:

- The system is complex
- There are several separate validation activities
- Multiple deployment sites are involved
- Multiple stakeholders have to be satisfied

There is also the question of how comprehensive to make the validation effort. It is impossible to validate all possible combinations of actions under all possible operational situations. A good rule of thumb is: if it was important enough to write down as a need or scenario, then it should be validated, at least once. This may not, for example, validate all possible failure mode conditions or all possible incident scenarios. In-process validation performed on the needs, will help ensure that end to end validation of the system will meet the stakeholder needs.

CHECKLIST: CRITICAL INFORMATION

- ✓ Does the Validation Plan answer all the questions of who, what, where, and when?
- ✓ Does the Validation Plan make clear what needs to happen if an unexpected situation or a failure is encountered?
- ✓ Does the Validation Plan document the configuration of the hardware and software?
- ✓ Are all applicable needs and scenarios traced to a validation case?

VALIDATION PLAN TEMPLATE

Section	Contents
Title Page	The title page should follow the Sponsoring agency's procedures or style guide. At a minimum, it should contain the following information: VALIDATION PLAN FOR THE (insert name of project) - (insert name of Sponsoring agency) Date that the document was formally approved The organization responsible for preparing the document Internal document control number, if available Revision version and date issued
1.0 Purpose of Document	This section identifies the type of validation activity to be performed within this Validation Plan. For instance, this activity may validate the entire system, a sub-system, the deployment at a site, a burn-in, or any other validation activity called for in the Program Plan or in the SEMP.
2.0 Scope of Project	This section gives a brief description of the planned project and the purpose of the system to be built. Special emphasis is placed on the project's user needs and issues that must be addressed and validated. This section also describes the environment in which the project operates. It identifies the organization structures that encompass all stakeholders. It also gives a brief description of the role to be played by each stakeholder. This includes ad hoc and existing management work groups and multi-disciplinary technical teams that should be formed to support the project.
3.0 Referenced Documents	This is a list of all documents used in the preparation of this Validation Plan. This almost always includes the Project Plan, the SEMP (if one was written), and the applicable Requirements Documents. Reference to other documents, such as descriptions of external systems, standards, a Concept of Operations, and manuals may also be included.

Section	Contents
4.0 Conducting Validation	This section provides details on how Validation is accomplished. It defines: who does the validation; when and where it is to be done; the responsibilities of each participant before, during, and after validation; the deployed hardware and software configuration; and the documents to be prepared as a record of the validation activity. This section defines how anomalies are to be handled (that is, what to do when an unexpected situation or a failure occurs during validation). In general, the following information should be included in this section: • A description of the participating organizations and personnel and identification of their roles and responsibilities. This may include for example, a validation conductor, validation recorder, operators, and/or engineering support. • Identification of the location of the validation effort, that is, the place, or places, where the validation must be observed. • The deployed hardware and software configuration for all of the validation cases, including hardware and software under validation and any supporting equipment, software, or external systems. Several configurations may be necessary. • Identification of the documents to be prepared to support the validation, including Validation Procedures, a Validation Report and descriptions of special equipment and software. • Details of the actual conduct of validation, including: • Notification of participants • Emphasis on the management role of the validation conductor • Procedures for approving last minute changes to the procedures • The processes for handling an unexpected situation or a failure, including recording of critical information, determination of whether to stop the validation, restart, or skip a procedure, resolution of the cause of a failure (e.g. fix the software, reset the system, repeat the procedure when conditions return to the planned state and/or change the requirements), and determination of the re-validation activities necessary as a result of the failure.

Section	Contents
5.0 Validation Identification	This section identifies the specific validation cases to be performed. A validation case is a logical grouping of functions and performance criteria (all from the Concept of Operations Documents) that are to be validated together. For instance, a specific validation case may cover all the control of traffic during the AM peak hour. There may be several individual objectives and associated performance measures that define this capability, and they all are validated in one case. The actual grouping of objectives into a case is arbitrary. They should be related and easily combined into a reasonable set of procedure actions. Each case should contain at least the following information: • A description name and a reference number • A complete list of needs and scenarios to be validated. For ease of tracing of needs and scenarios into the Validation Plan and other documents, the needs and scenarios are given numbers, so they can be accurately and conveniently referenced without repetition. • A description of the objective of the validation case, usually taken from the wording of the need or scenario, to aid the reader understanding the scope of the case • Any data to be recorded or noted during validation, such as expected results of a step. Other data, such as a recording of traffic volumes by the system, may be required to validate the performance of the system. • A statement of the pass/fail criteria. Often, this is just a statement that the system operates per the need or scenario. • A description of the validation configuration. That is a list of the hardware and software items needed for validation and how they should be connected. Often, the same configuration is used for several validation cases • A list of any other important assumptions and constraints necessary for conduct of the validation case

Appendix B
Concept of Operations
Table of Samples Statements

Con Ops Reference Number	Concept of Operations Sample Statements
1	**1 Chapter 1: Scope**
1.1	1.1 Document Purpose and Scope
1.1-1	The scope of this document covers the consideration of adaptive signal control technology (ASCT) for use within (describe the agency and/or geographic area covered by this consideration).
1.1-2	This document describes and provides a rationale for the expected operations of the proposed adaptive system.
1.1-3	It documents the outcome of stakeholder discussions and consensus building that has been undertaken to ensure that the system that is implemented is operationally feasible and has the support of stakeholders.
1.1-4	The intended audience of this document includes: system operators, administrators, decision-makers, elected officials, other nontechnical readers and other stakeholders who will share the operation of the system or be directly affected by it.
1.2	1.2 Project Purpose and Scope
1.2-1	An adaptive traffic signal system is one in which some or all the signal timing parameters are modified in response to changes in the traffic conditions, in real time.
1.2-2	The purpose of providing adaptive control in this area is to overcome (describe why it is needed, such as to overcome specific deficiencies or limitations in the existing system)
1.2-3	This project will add adaptive capabilities to the existing coordinated signal system.
1.2-4	This project will replace the existing coordinated traffic signal system to provide adaptive control.
1.2-5	All the capabilities of the existing coordinated system will be maintained.
1.2-6	The adaptive capability will be available at all signalized intersections within the agency's jurisdiction.
1.2-7	Adaptive capability will be provided for all coordinated signals within (describe the area to be covered)
1.2-8	The adaptive capability will be provided for signals operated by (name all the agencies whose signals will be part of the system)
1.2-9	Interfaces will be provided to the signal system operated by (name any agency whose signal system will be integrated or interfaced with the new adaptive system)
1.2-10	The adaptive system will be integrated with (name any other systems, such as an ICM or external decision support system)
1.3	1.3 Procurement
1.3.0-1	The ASCT system will be procured using (Edit this or choose alternative statement.)
1.3.0.1.0-1	a combination of best value procurement for software and system integration services, and low-bid procurement for equipment and construction services.
1.3.0.1.0-2	a best value procurement process based on responses to a request for proposals.
1.3.0.1.0-3	a low-bid process based on detailed plans and technical specifications.

Con Ops Reference Number	Concept of Operations Sample Statements
1.3.0-2	A request for qualifications (RFQ) will be issued to all potential vendors. Responses will be used to develop a short list of suitable systems and a request for proposals (RFP) will be issued to those vendors. The selected system will be the one that provides the best value, subject to financial and schedule constraints.
1.3.0-3	Field equipment (parts and labor) will be procured using a low-bid process based on detailed plans and technical specifications.
1.3.0-4	A detailed procurement plan will be prepared after the system requirements have been determined.
2	**2 Chapter 2: Referenced Documents**
2.0-1	The following documents have been used in the preparation of this Concept of Operations and stakeholder discussions. Some of these documents provide policy guidance for traffic signal operation in this area, some are standards with which the system must comply, while others report the conclusions of discussions, workshops and other research used to define the needs of the project and subsequently identify project requirements. References Specific to the Adaptive Locations - Business Planning / Strategic Planning Documents for relevant agencies - Concept of Operations for related agency/facility-specific systems - Requirements of related systems - Studies identifying operational needs - Regional ITS Architecture documents - Planning studies and Master Plans - Transportation Improvement Programs (TIP) - Long Range Transportation Plans
2.0-1.0-1	Systems Engineering
2.0-1.0-2	- "Systems Engineering Guidebook for ITS", California Department of Transportation, Division of Research & Innovation, Version 3.0, <http://www.fhwa.dot.gov/cadiv/segb/> - "Systems Engineering for Intelligent Transportation Systems, An Introduction for Transportation Professionals", <http://ops.fhwa.dot.gov/publications/seitsguide/index.htm> - "Developing Functional Requirements for ITS Projects", Mitretek Systems, April 2002 - "Developing and Using a Concept of Operations in Transportation Management System, FHWA TMC Pooled-Fund Study (http://tmcpfs.ops.fhwa.dot.gov/cfprojects/new_detail.cfm?id=38&new=0) - NCHRP Synthesis 307: Systems Engineering Processes for Developing Traffic Signal Systems
	Adaptive Signals
2.0-1.0-3	* NCHRP Synthesis 403: "Adaptive Traffic Control Systems: Domestic and Foreign State of Practice" (http://onlinepubs.trb.org/onlinepubs/nchrp/nchrp_syn_403.pdf)

Con Ops Reference Number	Concept of Operations Sample Statements
	ITS, Operations, Architecture, Other
2.0.1.0-4	• FHWA Rule 940, Federal Register / Vol. 66, No. 5 / Monday, January 8, 2001 / Rules and Regulations, DEPARTMENT OF TRANSPORTATION, Federal Highway Administration 23 CFR Parts 655 and 940, [FHWA Docket No. FHWA-99-5899] RIN 2125-AE65 Intelligent Transportation System Architecture and Standards • Regional ITS Architecture Guidance Document; "Developing, Using, and Maintaining an ITS Architecture for your Region; National ITS Architecture Team; October, 2001
	NTCIP
2.0.1.0-5	• List applicable NTCIP standards ADD MORE COMPLETE LIST HERE SO USERS CAN PICK AND CHOOSE.
	NEMA
2.0.1.0-6	• List applicable NEMA standards INSERT MORE COMPLETE LIST SO USER CAN PICK AND CHOOSE
	PROCUREMENT
2.0.1.0-7	• NCHRP 560: <http://onlinepubs.trb.org/onlinepubs/nchrp/nchrp_rpt_560.pdf> • Special Experimental Project 14 (SEP 14): <http://www.fhwa.dot.gov/programadmin/contracts/sep_a.cfm> • The Road to Successful ITS Software Acquisition:<http://www.fhwa.dot.gov/publications/research/operations/its/98036/rdsuccessvol2.pdf>
3	**3 Chapter 3: User-Oriented Operational Description**
3.1	3.1 The Existing Situation
3.1.1	*3.1.1 Network Characteristics*
3.1.1.1	3.1.1.1 Arterial
3.1.1.1.0-1	The arterial has regularly spaced signalized intersections. The spacing between major intersections is approximately XX, with less important intersections spaced at XX. The locations at which ASCT is being considered are illustrated in FIGURE XX.
3.1.1.1.0-2	The free-flow travel time between major intersections is approximately XX seconds. (Expand this description as appropriate to cover additional arterials or networks.)
3.1.1.1.0-3	The travel time between key intersections allows two-way progression when cycle lengths of XX seconds (CL=travel time) or YY (CL= 2x travel time) seconds can be used. (Add descriptions of additional cycle lengths if appropriate.)
3.1.1.1.0-4	The arterial has irregularly spaced signalized intersections, and there is no "natural" cycle length that allows two-way progression.

Con Ops Reference Number	Concept of Operations Sample Statements
3.1.1.1.0-5	During the peak periods, the cycle length is generally determined by the needs of one or more critical intersections.
3.1.1.1.0-6	The cycle length required to service traffic at the critical intersection(s) is generally close to a "natural" cycle length.
3.1.1.1.0-7	The capacity of the arterial changes during the day, with parking restrictions providing higher capacity during peak periods.
3.1.1.2	3.1.1.2 Grid
3.1.1.2.0-1	The network is a uniform grid. (Expand this description and include figures as appropriate.)
3.1.1.2.0-2	The signal phasing is similar at all intersections, and is typically... (Describe the phasing)
3.1.1.2.0-3	Several intersections have multi-phase intersections that require a higher cycle length than most intersections.
3.1.1.2.0-4	Several roads in the grid are higher capacity arterials.
3.1.1.2.0-5	The signals along one (or more) higher capacity street(s) (specify if appropriate) generally require a higher cycle length than most of the grid intersections.
3.1.1.2.0-6	The capacity of some roads in the network changes during the day, with parking restrictions providing higher capacity during peak periods.
3.1.1.3	3.1.1.3 Isolated Intersection or Small Group
3.1.1.3.0-1	There is one critical intersection in the project area, and the timing of adjacent intersections mainly needs to accommodate progression for the platoons serviced by the critical intersection.
3.1.1.3.0-2	The system will be used to improve operation at a single, isolated intersection that does not operate efficiently with typical vehicle-actuated operation. It requires (choose as appropriate): • Different phase sequences at different times of the day • Phase reserve to prevent queue overflow in turn bays • Different cycle lengths for different periods • Different splits (phase maximums) for different periods
3.1.1.4	3.1.1.4 Freeway Interchange
3.1.1.4.0-1	The project location has several closely spaced intersections with major turning movements at a freeway interchange. It requires careful management of queue lengths on some approaches.
3.1.1.4.0-2	Queuing from on-ramps affects the distribution of traffic across the lanes on the arterial.
3.1.1.4.0-3	Queuing from on-ramps affects the saturation flow of some movements during green.
3.1.1.5	3.1.1.5 Jurisdictions
3.1.1.5.0-1	The signals are owned and/or operated and/or maintained by several separate agencies. (Describe which signals are owned, operated and maintained by which agency. Refer to any relevant MOU's, and service and maintenance agreements.)

Con Ops Reference Number	Concept of Operations Sample Statements
3.1.2	*3.1.2 Traffic Characteristics*
3.1.2.1	3.1.2.1 Overview
3.1.2.1.0-1	The traffic characteristics are illustrated in FIGURE XX. (Include graphs and figures with an appropriate explanation.)
3.1.2.2	3.1.2.2 Peak Periods
3.1.2.2.0-1	There are heavily directional commuter peaks. E.g., during the AM peak, traffic is heavily directional in the XX direction. The peak hour volume in the XX direction is xxxx, while the peak hour volume in the YY direction is yyyy.
3.1.2.2.0-2	Traffic is balanced during commuter peaks. E.g., during the AM peak, the volumes in the two directions are similar, with xxxx vehicles per hour in the XX direction and yyyy vehicles per hour in the YY direction.
3.1.2.2.0-3	Traffic conditions vary during the commuter peaks. E.g., During early part of the AM peak period, traffic flows predominantly in the XX direction, then later in the period it becomes balanced (or flows predominantly in the opposite direction). Choose description as appropriate.
3.1.2.2.0-4	A major traffic generator close to the intersections to be coordinated has non-cyclical peaks. (E.g., a shipping port at which the peak traffic is influenced by the tides, or a military base at which activity does not operate on a weekly cycle.) The direction and magnitude of the peak hour flows is unpredictable and non-uniform.
3.1.2.3	3.1.2.3 Business Hours
3.1.2.3.0-1	Business hours volumes are light between the peaks
3.1.2.3.0-2	Business hours volumes in the two directions are balanced between the peaks.
3.1.2.3.0-3	Business hours flows are predominantly in the XX direction.
3.1.2.3.0-4	Business hours flows are directional, but vary during the day. E.g., during the morning business hours, the predominant flow is in the XX direction, while during the afternoon hours it is in the YY direction.
3.1.2.3.0-5	During the lunchtime period, there are minor peaks, as illustrated in FIGURE XX. (Include graphics as appropriate.)
3.1.2.4	3.1.2.4 Evenings
3.1.2.4.0-1	During the evenings after the PM peak, the flows are....(Select an appropriate description and illustrate with graphics as appropriate)
3.1.2.4.0-1.0-1	Directional
3.1.2.4.0-1.0-2	Balanced
3.1.2.4.0-1.0-3	Heavy
3.1.2.4.0-1.0-4	Light
3.1.2.5	3.1.2.5 Weekends

Con Ops Reference Number	Concept of Operations Sample Statements
3.1.2.5.0-1	During the weekends, the flows are....(Select an appropriate description and illustrate with graphics as appropriate)
3.1.2.5.0-1.0-1	Balanced weekend flows
3.1.2.5.0-1.0-2	Changing weekend patterns
3.1.2.5.0-1.0-3	Saturday or Sunday peaks (Related to retail, recreation, worship and other factors.)
3.1.2.5.0-1.0-4	Weekend retail traffic
3.1.2.5.0-1.0-5	Weekend recreational traffic
3.1.2.6	3.1.2.6 Events and Incidents
3.1.2.6.0-1	Heavily directional event traffic is experienced in this area. (give details such as time of events, duration, day of week, volumes)
3.1.2.6.0-2	Heavily directional incident-related traffic is experienced in this area. (E.g., during peak periods, during off-peak period, during weekends)
3.1.2.7	3.1.2.7 General
3.1.2.7.0-1	There is a high proportion of turning traffic along the arterial or within the network.
3.1.2.7.0-2	At (some or most or all) intersections there is a high proportion of turning traffic.
3.1.2.7.0-3	Queues often overflow from turn bays at (locations) during (periods of time).
3.1.2.7.0-4	Traffic along the arterial is predominantly through traffic.
3.1.2.7.0-5	The origin or destination of most traffic lies within the corridor or grid.
3.1.2.7.0-6	There are significant turning movements onto and off the coordinated route. (Expand the description as appropriate.)
3.1.2.7.0-7	Traffic conditions change quickly when... (describe the circumstances)
3.1.2.8	3.1.2.8 Future Traffic Conditions
3.1.2.8.0-1	Describe any changes in traffic conditions that are expected to occur within the likely expected life of the proposed ASCT.
3.1.3	3.1.3 Signal Grouping
3.1.3.0-1	The locations of signals to be operated under adaptive control are illustrated in figure XX. (Include appropriate figures, describe future needs e.g. adaptive control may be expanded to a significant percentage of the signals within the jurisdiction.)
3.1.3.0-2	All the signals are relatively close and are expected to be coordinated as one group.
3.1.3.0-3	While the signals are relatively close, the traffic conditions are such that they will normally be coordinated as two (or more) separate and independent groups.

Con Ops Reference Number	Concept of Operations Sample Statements
3.1.3.0-4	While the signals are relatively close, the traffic conditions vary and sometimes they would be expected to be coordinated as one group, while at other times they may be coordinated as two (or more) separate and independent groups.
3.1.3.0-5	Although the signals are all on the one route, the distance between them is sufficiently great that they will normally be coordinated as two (or more) separate and independent groups.
3.1.3.0-6	The signals to be coordinated are on two (or more) arterials separated by XX miles, and will always be operated as two separate groups.
3.1.3.0-7	While the signals will normally be operated as two (or more) separate and independent groups, there are occasions (such as when there is a major incident on the parallel freeway) when they should operate as one coordinated unit.
3.1.4	*3.1.4 Land Use Characteristics*
3.1.4.1	3.1.4.1 Existing Land Uses
3.1.4.1.0-1	(Edit or or select from the following statements if your situation includes an arterial.) The arterial….
3.1.4.1.0-1.0-1	Passes through residential neighborhoods.
3.1.4.1.0-1.0-2	Frontage land uses are mainly retail (E.g., strip mall with numerous driveways, shopping mall with several signalized driveways, big box outlet).
3.1.4.1.0-1.0-3	Frontage land uses are mainly offices.
3.1.4.1.0-1.0-4	Frontage land uses are mainly commercial.
3.1.4.1.0-1.0-5	Frontage land uses are mainly service trades.
3.1.4.1.0-1.0-6	Frontage land uses are mainly manufacturing.
3.1.4.1.0-1.0-7	Serves a mixture of land uses, including (delete those not applicable) residential, office, commercial, retail, service trades, manufacturing, education (specify high school, elementary school, junior college, university, etc.).
3.1.4.1.0-1.0-8	Serves a major event center. (specify stadium, park/open space, market, etc.)
3.1.4.1.0-1.0-9	Provides a parallel route to a freeway.
3.1.4.1.0-1.0-10	Provides access to a freeway interchange.
3.1.4.1.0-2	(Edit or or select from the following statements if your situation includes a grid or similar network.) The road network…
3.1.4.1.02.0-1	Encompasses a downtown area with mixed uses and a variety of activities. (Edit the description as appropriate.)
3.1.4.2	3.1.4.2 Future Land Use Changes

Con Ops Reference Number	Concept of Operations Sample Statements
3.1.4.2.0-1	Describe any changes in land use that are expected to occur within the likely expected life of the proposed ASCT.
3.1.4.3	3.1.4.3 Pedestrians and Public Transit
3.1.4.3.0-1	This section describes the influence of pedestrians on the signal operation.
3.1.4.3.0-1.0-1	Pedestrian delays are a factor in choosing phasing and timing parameters.
3.1.4.3.0-1.0-2	Pedestrians impede turning movements at… (Describe the locations.)
3.1.4.3.0-1.0-3	Pedestrians are present every cycle.
3.1.4.3.0-1.0-4	Pedestrians are present most cycles.
3.1.4.3.0-1.0-5	Pedestrian phases are rarely called.
3.1.4.3.0-2	This section describes the influence of transit on the signal operation.
3.1.4.3.0-2.0-1	There are XX bus lines operating along the route (or within the network). The buses operate at a frequency of XX per hour during peak periods, and (describe operation during other periods).
3.1.4.3.0-2.0-2	Buses enter (and/or leave, and/or cross) the coordinated route at (describe the locations where buses turn or cross coordinated routes).
3.1.4.3.0-2.0-3	A light rail line operates along the coordinated route. (Describe the operation, such as shared lanes, exclusive lanes, in the median, side running, with or without signal priority)
3.1.4.3.0-2.0-4	A light rail line operates parallel to the coordinated route in a separate right of way. It crosses each of the cross streets approximately XX feet from the intersection, and its operation preempts the signals on the coordinated route. (Describe how it actually operates in your situation.)
3.1.4.4	3.1.4.4 Agencies
3.1.4.4.0-1	The existing signal system is operated by (name the agency). Some intersections are controlled by signals belonging to (name other agencies). These are controlled by XX agency (or coordinated with the system operated by YY agency).
3.1.4.4.0-2	The effectiveness of (name to agency, such as transit and fire department) is affected by the operation of the signal system.
3.1.4.5	3.1.4.5 Existing Architecture
3.1.4.5.0-1	The existing system architecture is illustrated in FIGURE XX. (Provide an appropriate system network block diagram, and describe the following elements, as applicable.)
3.1.4.5.0-1.0-1	TMC and workstations
3.1.4.5.0-1.0-2	Local hubs, on-street masters, etc.
3.1.4.5.0-1.0-3	Communications infrastructure (e.g., fiber optic cable, twisted wire pair cable, serial or Ethernet communications)
3.1.4.5.0-1.0-4	Detection locations and technology (e.g., video, loops or other technology; stop line, advance or mid-block detection zones)

Con Ops Reference Number	Concept of Operations Sample Statements
3.2	**3.2 Limitations of the Existing system**
	The following statements summarize the limitations of the existing system that prevent it from satisfactorily accommodating the traffic situations described above. (Select from the following samples and create new descriptions that fit your situation.)
3.2.0-1	The existing system cannot recognize the onset of peak periods, so the peak period coordination plan introduction times are set conservatively to ensure they cover the normal variation in duration and intensity of the peak. This means that the timing is often less efficient during the early and late parts of the peak periods.
3.2.0-2	The peak direction fluctuates during the peak, so the peak period plan is a compromise. An adaptive system would be expected to recognize the direction of heaviest flow in real time and react accordingly, rather than use a plan that is less efficient but can accommodate a range of flows.
3.2.0-3	The coordinated signal operation is often disrupted by light rail priority (or rail preemption, or bus signal priority). An adaptive system may be expected to recover from these disruptions more quickly than the existing system. (Describe how the existing system recovers and how long it takes. Quantify how much improvement may be expected from an adaptive system.)
3.2.0-4	The existing system cannot detect unexpected changes in traffic demand as a result of incidents on the adjacent freeway. As a result, the congestion on the arterials is greater than would be the case if the signal timing could automatically adjust to the unexpected conditions. This would also reduce the need for manual intervention by operators when the incident is brought to their attention.
3.2.0-5	The existing system cannot detect the changes in traffic conditions before and after games at the XXXX stadium. As a result, the coordination plan introduction times are set very conservatively, and they generally begin operating before they are needed and continue until well after the traffic disperses. An adaptive system could be expected to reduce this inefficiency and match the signal timing more closely to the actual traffic patterns.
3.2.0-6	
3.3	**3.3 Proposed Improvements to the System**
	This section describes in broad terms the improvements that are desirable in order to address the limitations described above. The main improvements that are desired are: (Select from the samples below and create new descriptions that suit your situation.)
3.3.0-1	• Recognize changes in traffic conditions and react quickly and automatically to accommodate those changes.
3.3.0-2	• Overcome the institutional boundaries that currently prevent the signals under the control of the different jurisdictions from operating in a coordinated fashion.
3.3.0-3	• More efficiently accommodate rail, emergency vehicles and transit vehicles and more quickly recover from preemption and priority.
3.3.0-4	• Improve the management of queues within the network.
3.3.0-5	• Recognize the existence of differing traffic conditions in various parts of the network and react in each section appropriately.
3.3.0-6	

Con Ops Reference Number	Concept of Operations Sample Statements
3.3.0-7	• Improve the productivity of staff by automating many of the routine processes.
3.3.0-8	• Keep signal timing current rather than letting its efficiency deteriorate between periodic signal re-timing efforts.
3.4	3.4 Vision, Goals and Objectives for the Proposed System
3.4.1	3.4.1 Vision
3.4.1-1	The vision of the ASCT system is to provide an advanced traffic control system that responds to changing traffic conditions, and reduces delays and corridor travel times, while balancing multimodal transportation needs. (Customize this statement to suit your situation.)
3.4.2	3.4.2 Goals
3.4.2-1	The goals of the ASCT system are: (Select from the following items and customize to suit your situation.)
3.4.2-1.0-1	• Support vehicle, pedestrian and transit traffic mobility.
3.4.2-1.0-2	• Provide measurable improvements in personal mobility
3.4.2-1.0-3	• Support interoperability between agencies
3.4.2-1.0-4	• Support regional systems
3.4.2-1.0-5	• Support congestion and environment policy objectives
3.4.2-1.0-6	• Meet a timely project implementation schedule
3.4.3	3.4.3 User Objectives
3.4.3.0-1	The objectives of the adaptive system that support the stated goals are: (Select from the following items and customize to suite your situation.) To support vehicle, pedestrian and transit traffic mobility:
3.4.3.0-1.0-1	• Be capable of supporting priority operations for light rail and buses • Allow effective use of all controller features currently in use or proposed to be used • Minimize adverse effects caused by preemption and unexpected events To support measurable improvements in personal mobility:
3.4.3.0-1.0-2	• Adjust operations to changing conditions • Reduce delays • Reduce travel times • Provide the same level of safety provided by the existing system to vehicles, pedestrians and transit. To support agency interoperability:
3.4.3.0-1.0-3	• Provide facilities for data exchange and control between systems • Allow remote monitoring and control • Adhere to applicable traffic signal and ITS design standards

Con Ops Reference Number	Concept of Operations Sample Statements
	To support regional systems:
3.4.3.0.1.0-4	<ul><li>Be compliant with the regional ITS architecture</li><li>Allow center-to-center and system-to-system communication</li><li>Connect to regional traffic control systems</li><li>Report traffic conditions to regional traffic conditions information systems</li></ul>
	To support environmental objectives:
3.4.3.0.1.0-5	<ul><li>Reduce vehicle emissions through improvements in appropriate determinants such as vehicle stops and delays</li></ul>
	To support a timely schedule:
3.4.3.0.1.0-6	<ul><li>Be sufficiently mature and robust that risk is low and little or no development time will be required.</li><li>Be ready for full operation by (specify an appropriate date if you have an imposed deadline)</li></ul>
3.4.4	3.4.4 Operational Objectives
3.4.4.0.1	The operational objectives of the ASCT system will be to: (Select the samples appropriate to your situation)
3.4.4.0.1.0-1	Smooth the flow of traffic along coordinated routes
3.4.4.0.1.0-2	Maximize the throughput of traffic along coordinated routes
3.4.4.0.1.0-3	Equitably serve adjacent land uses
3.4.4.0.1.0-4	Manage queues, to prevent excessive queuing from reducing efficiency
3.4.4.0.1.0-5	Control operation using a combination of these objectives
3.4.4.0.1.0-6	Control operation by changing the objectives under various circumstances
3.4.4.0.1.0-7	For a critical isolated intersection, maximize intersection efficiency.
3.5	3.5 Strategies to be Applied by the Improved System
3.5.0-1	The adaptive coordination and control strategies that may be employed to achieve the operational objectives are: (Select the samples that are applicable to your situation)
3.5.0.1.0-1	<ul><li>Provide a pipeline along a coordinated route to maximize the throughput during periods of high demand;</li></ul>
3.5.0.1.0-2	<ul><li>Provide a pipeline along a coordinated route to smooth the flow of traffic in one or both directions;</li></ul>
3.5.0.1.0-3	<ul><li>Distribute phase times in a way that equitably shares the green time between various movements and minimizes the risk of phase failures;</li></ul>
3.5.0.1.0-4	<ul><li>Manage queues so they do not exceed the available storage capacity and are located so they do not affect the capacity of other movements;</li></ul>

Con Ops Reference Number	Concept of Operations Sample Statements
3.5.0-1.0-5	• Manage the distribution of green times for vehicles and pedestrians in an equitable manner;
3.5.0-1.0-6	• Employ a combination of these strategies when they are compatible.
3.5.0-1.0-7	Not Used
3.6	*3.6 Alternative Non-Adaptive Strategies Considered*
3.6.1	*3.6.1 Traffic Responsive Pattern Selection*
3.6.1.0-1	TRPS has been operated in the past. It has been successful, but has some limitations that affect its effectiveness. (Explain what limitations are evident.) OR It has not been successful for the following reasons (Explain your experience with TRPS)
3.6.1.0.2	Could TRPS operation be used? (If not, why not)
3.6.1.0.3	How successful would TRPS be if it were used.
3.6.2	*3.6.2 Complex Coordination Features*
	The following features are currently used in coordination patterns. These features will need to remain available in fallback operation should the ASCT fail. (Select from the list as appropriate.)
3.6.2.0-1	• Multiple (repeat) phases or phase reservice • Variable phase sequence • Omit phase under some circumstances • Detector switching • Coordinate different phases at different times • Coordinate turning movement phases • Coordinate beginning or end of green • Early release of hold • Hold the position of uncoordinated phases • Late phase introduction • Stop-in-walk • Dynamic max • Double cycle or half cycle

Con Ops Reference Number	Concept of Operations Sample Statements
3.6.2.0-2	The following features have not been used in the current coordination patterns. While they have been considered, they are not suitable in this situation for the following reasons. (Select from the list as appropriate, and explain why each is not suitable.) • Multiple (repeat) phases or phase reservice • Variable phase sequence • Omit phase under some circumstances • Detector switching • Coordinate different phases at different times • Coordinate turning movement phases • Coordinate beginning or end of green • Early release of hold • Hold the position of uncoordinated phases • Late phase introduction • Stop-in-walk • Dynamic max • Double cycle or half cycle

Con Ops Reference Number	Concept of Operations Sample Statements	System Requirements (Tailor as required - See Guidance)	Guidance Section
4	**4 Chapter 4: Operational Needs**		
4.0-1	This chapter describes the operational needs of the users that should be satisfied by the proposed ASCT system. Each of these statements describes something that the system operators need to be able to achieve. Each of these needs will be satisfied by compliance with one or more system requirements. In the attached list of requirements, each one is linked to one or more of these needs statements.		
4.1	**4.1 Adaptive Strategies**		
4.1.0-1	The system operator needs the ability to implement different strategies individually or in combination to suit different prevailing traffic conditions. These strategies include:		3.4 3.5
4.1.0.1.0-1	• Maximize the throughput on coordinated routes Note to user when selecting these requirements: Select from requirements in the 2.2 group when sequence-based systems are allowed (sequence-based systems explicitly calculate cycle, offset, and split) Select from requirements in the 2.3 group when non-sequence-based systems are allowed (non-sequence-based systems do not explicitly calculate cycle, offset, and split). (Select requirements from both groups when the vendor is given the choice of supplying one type of adaptive operation or the other.)	2.2.0-4 (Sequence-based only) The ASCT shall calculate offsets to suit the current coordination strategy for the user-specified reference point for each signal controller along a coordinated route within a group. 2.2.0-4.0-1 (Sequence-based only) The ASCT shall apply offsets for the user-specified reference point of each signal controller along a coordinated route. 2.1.1.0-7.0-1 When current measured traffic conditions meet user-specified criteria, the ASCT shall alter the state of the signal controllers, maximizing the throughput of the coordinated route. 2.2.0-5.0-3 (Sequence-based only) The ASCT shall calculate optimum cycle length according to the user-specified coordination strategy.	

Con Ops Reference Number	Concept of Operations Sample Statements	System Requirements (Tailor as required - See Guidance)	Guidance Section
		2.2.0-5 (Sequence-based only) The ASCT shall calculate a cycle length for each cycle based on its optimization objectives (as required elsewhere, e.g., progression, queue management, equitable distribution of green).	
		2.3.0-3 (Non-sequence-based only) At non-critical intersections within a group, the ASCT shall calculate the time at which a user-specified phase shall be green, relative to a reference point at the critical intersection, to suit the current coordination strategy.	
		2.3.0-2 (Non-sequence-based only) The ASCT shall calculate the appropriate state of the signal to suit the current coordination strategy at the critical signal controller. (A critical signal controller is defined by the user.)	
		2.3.0-4 (Non-sequence-based only) When demand is present, the ASCT shall implement a user-specified maximum time between successive displays of each phase at each intersection.	
		2.1.1.0-7 The ASCT shall alter the adaptive operation to achieve required objectives in user-specified conditions. (The required objectives are specified in Needs Statement 4.1.0-1. Responding to this requirement demonstrates how the proposed system allows the user to define the conditions at which the objectives shift and their associated requirements are fulfilled.) (The alteration may be made by adjusting parameters or by directly controlling the state of signal controllers.)	
		2.2.0-2 (Sequence-based only) The ASCT shall select cycle length based on a time of day schedule.	

Con Ops Reference Number	Concept of Operations Sample Statements	System Requirements (Tailor as required - See Guidance)	Guidance Section
		2.2.0-5.0-1 (Sequence-based only) The ASCT shall limit cycle lengths to user-specified values.	
		2.2.0-5.0-2 (Sequence-based only) The ASCT shall limit cycle lengths to a user-specified range.	
		2.2.0-5.0.4 (Sequence-based only) The ASCT shall limit changes in cycle length to not exceed a user-specified value.	
		2.2.0-5.0.4.0-1 (Sequence-based only) The ASCT shall increase the limit for the following XX cycles based on a change in conditions.	
		2.2.0-5.0.4.0.1.0-2 (Sequence-based only) The increased limit shall be user-defined.	
		2.2.0-5.0.4.0.1.0-1 (Sequence-based only) The change in conditions shall be defined by XX successive adaptive increases in cycle length at the maximum rate.	
		2.1.1.0-10 The ASCT shall determine the order of phases at a user-specified intersection. (The calculation will be based on the optimization function.)	

Con Ops Reference Number	Concept of Operations Sample Statements	System Requirements (Tailor as required - See Guidance)	Guidance Section
4.1.0.1.02	• Provide smooth flow along coordinated routes Note to user when selecting these requirements: Select from requirements in the 2.2 group when sequence-based systems are allowed (sequence-based systems explicitly calculate cycle, offset, and split). Select from requirements in the 2.3 group when non-sequence-based systems are allowed (non-sequence-based systems do not explicitly calculate cycle, offset, and split). (Select requirements from both groups when the vendor is given the choice of supplying one type of adaptive operation or the other.)	2.2.0-4 (Sequence-based only) The ASCT shall calculate offsets to suit the current coordination strategy for the user-specified reference point for each signal controller along a coordinated route within a group. 2.2.0-4.0.1 (Sequence-based only) The ASCT shall apply offsets for the user-specified reference point of each signal controller along a coordinated route. 2.1.1.0.7.0-4 When current measured traffic conditions meet user-defined criteria, the ASCT shall alter the state of signal controllers providing two-way progression on a coordinated route 2.2.0-5.0-3 (Sequence-based only) The ASCT shall calculate optimum cycle length according to the user-specified coordination strategy. 2.2.0-5 (Sequence-based only) The ASCT shall calculate a cycle length for each cycle based on its optimization objectives (as required elsewhere, e.g., progression, queue management, equitable distribution of green). 2.3.0-3 (Non-sequence-based only) At non-critical intersections within a group, the ASCT shall calculate the time at which a user-specified phase shall be green, relative to a reference point at the critical intersection, to suit the current coordination strategy. 2.3.0-2 (Non-sequence-based only) The ASCT shall calculate the appropriate state of the signal to suit the current coordination strategy at the critical signal controller. (A critical signal controller is defined by the user.)	

Con Ops Reference Number	Concept of Operations Sample Statements	System Requirements (Tailor as required - See Guidance)	Guidance Section
		2.3.0-4 (Non-sequence-based only) When demand is present, the ASCT shall implement a user-specified maximum time between successive displays of each phase at each intersection.	
		2.2.0-2 (Sequence-based only) The ASCT shall select cycle length based on a time of day schedule.	
		2.2.0-5.0-1 (Sequence-based only) The ASCT shall limit cycle lengths to user-specified values.	
		2.2.0-5.0-2 (Sequence-based only) The ASCT shall limit cycle lengths to a user-specified range.	
		2.2.0-5.0-4 (Sequence-based only) The ASCT shall limit changes in cycle length to not exceed a user-specified value.	
		2.2.0-5.0-4.0-1 (Sequence-based only) The ASCT shall increase the limit for the following XX cycles based on a change in conditions.	
		2.2.0-5.0-4.0-1.0-2 (Sequence-based only) The increased limit shall be user-defined.	
		2.2.0-5.0-4.0-1.0-1 (Sequence-based only) The change in conditions shall be defined by XX successive adaptive increases in cycle length at the maximum rate.	
		2.1.1.0-10 The ASCT shall determine the order of phases at a user-specified intersection. (The calculation will be based on the optimization function.)	

Con Ops Reference Number	Concept of Operations Sample Statements	System Requirements (Tailor as required - See Guidance)	Guidance Section
4.1.0-1.0-3	• Distribute phase times in an equitable fashion Note to user when selecting these requirements: Select from requirements in the 2.2 group when sequence-based systems are allowed (sequence-based systems explicitly calculate cycle, offset, and split). Select from requirements in the 2.3 group when non-sequence-based systems are allowed (non-sequence-based systems do not explicitly calculate cycle, offset, and split). (Select requirements from both groups when the vendor is given the choice of supplying one type of adaptive operation or the other.)	2.1.1.0-7.0-3 When current measured traffic conditions meet user-specified criteria, the ASCT shall alter the state of signal controllers providing equitable distribution of green times. 2.2.0-3 (Sequence-based only) The ASCT shall calculate phase lengths for all phases at each signal controller to suit the current coordination strategy. 2.2.0-5.0-3 (Sequence-based only) The ASCT shall calculate optimum cycle length according to the user-specified coordination strategy. 2.2.0-5 (Sequence-based only) The ASCT shall calculate a cycle length for each cycle based on its optimization objectives (as required elsewhere, e.g., progression, queue management, equitable distribution of green). 2.4.0-3 The ASCT shall calculate optimum phase lengths, based on current measured traffic conditions. (The calculation is based on the optimization objectives.) 2.3.0-3 (Non-sequence-based only) At non-critical intersections within a group, the ASCT shall calculate the time at which a user-specified phase shall be green, relative to a reference point at the critical intersection, to suit the current coordination strategy. 2.3.0-2 (Non-sequence-based only) The ASCT shall calculate the appropriate state of the signal to suit the current coordination strategy at the critical signal controller. (A critical signal controller is defined by the user.)	3.4 3.5

Con Ops Reference Number	Concept of Operations Sample Statements	System Requirements (Tailor as required - See Guidance)	Guidance Section
		2.3.0-4 (Non-sequence-based only) When demand is present, the ASCT shall implement a user-specified maximum time between successive displays of each phase at each intersection.	
		2.1.1.0-7 The ASCT shall alter the adaptive operation to achieve required objectives in user-specified conditions. (The required objectives are specified in Needs Statement 4.1.0-1. Responding to this requirement demonstrates how the proposed system allows the user to define the conditions at which the objectives shift and their associated requirements are fulfilled.) (The alteration may be made by adjusting parameters or by directly controlling the state of signal controllers.)	
		2.2.0-2 (Sequence-based only) The ASCT shall select cycle length based on a time of day schedule.	
		2.1.1.0-8.0-1 The ASCT shall provide a user-specified maximum value for each phase at each signal controller.	
		2.1.1.0-8.0-1.0-1 The ASCT shall not provide a phase length longer that the maximum value.	
		2.1.1.0-8.0-2 The ASCT shall provide a user-specified minimum value for each phase at each signal controller.	
		2.1.1.0-8.0-2.0-1 The ASCT shall not provide a phase length shorter than the minimum value.	
		2.2.0-5.0-1 (Sequence-based only) The ASCT shall limit cycle lengths to user-specified values.	

Con Ops Reference Number	Concept of Operations Sample Statements	System Requirements (Tailor as required - See Guidance)	Guidance Section
		2.2.0-5.0.2 (Sequence-based only) The ASCT shall limit cycle lengths to a user-specified range.	
		2.2.0-5.0.4 (Sequence-based only) The ASCT shall limit changes in cycle length to not exceed a user-specified value.	
		2.2.0-5.0.4.0-1 (Sequence-based only) The ASCT shall increase the limit for the following XX cycles based on a change in conditions.	
		2.2.0-5.0.4.0.1.0-2 (Sequence-based only) The increased limit shall be user-defined.	
		2.2.0-5.0.4.0.1.0-1 (Sequence-based only) The change in conditions shall be defined by XX successive adaptive increases in cycle length at the maximum rate.	
		2.1.1.0-8 The ASCT shall provide maximum and minimum phase times.	
		2.4.0-3.0-1 The ASCT shall limit the difference between the length of a given phase and the length of the same phase during its next service to a user-specified value.	
		2.4.0-3.0-2 When queues are detected at user-specified locations, the ASCT shall execute user-specified timing plan/operational mode.	

Con Ops Reference Number	Concept of Operations Sample Statements	System Requirements (Tailor as required - See Guidance)	Guidance Section
4.1.0.1.0-4	• Manage the length of queues Note to user when selecting these requirements: Select from requirements in the 2.2 group when sequence-based systems are allowed (sequence-based systems explicitly calculate cycle, offset, and split). Select from requirements in the 2.3 group when non-sequence-based systems are allowed (non-sequence-based systems do not explicitly calculate cycle, offset, and split). (Select requirements from both groups when the vendor is given the choice of supplying one type of adaptive operation or the other.)	2.1.3.0-2 When queues are detected at user-specified locations, the ASCT shall execute user-specified timing plan/operational mode. 2.2.0-4 (Sequence-based only) The ASCT shall calculate offsets to suit the current coordination strategy for the user-specified reference point for each signal controller along a coordinated route within a group. 2.2.0-4.0-1 (Sequence-based only) The ASCT shall apply offsets for the user-specified reference point of each signal controller along a coordinated route. 2.1.1.0.7.0-2 When current measured traffic conditions meet user-specified criteria, the ASCT shall alter the state of signal controllers, preventing queues from exceeding the storage capacity at user-specified locations. 2.2.0-5.0-3 (Sequence-based only) The ASCT shall calculate optimum cycle length according to the user-specified coordination strategy. 2.2.0-5 (Sequence-based only) The ASCT shall calculate a cycle length for each cycle based on its optimization objectives (as required elsewhere, e.g., progression, queue management, equitable distribution of green). 2.3.0-3 (Non-sequence-based only) At non-critical intersections within a group, the ASCT shall calculate the time at which a user-specified phase shall be green, relative to a reference point at the critical intersection, to suit the current coordination strategy.	3.4 3.5

Con Ops Reference Number	Concept of Operations Sample Statements	System Requirements (Tailor as required - See Guidance)	Guidance Section
		2.1.3.0-1 The ASCT shall detect the presence of queues at pre-configured locations.	
		2.3.0-2 (Non-sequence-based only) The ASCT shall calculate the appropriate state of the signal to suit the current coordination strategy at the critical signal controller. (A critical signal controller is defined by the user.)	
		2.3.0-4 (Non-sequence-based only) When demand is present, the ASCT shall implement a user-specified maximum time between successive displays of each phase at each intersection.	
		2.2.0-2 (Sequence-based only) The ASCT shall select cycle length based on a time of day schedule.	
		2.1.3.0-3 When queues are detected at user-specified locations, the ASCT shall execute user-specified adaptive operation strategy.	
		2.1.3.0-4 When queues are detected at user-specified locations, the ASCT shall omit a user-specified phase at a user-specified signal controller.	
		2.2.0-5.0-1 (Sequence-based only) The ASCT shall limit cycle lengths to user-specified values.	
		2.2.0-5.0-2 (Sequence-based only) The ASCT shall limit cycle lengths to a user-specified range.	
		2.2.0-5.0-4 (Sequence-based only) The ASCT shall limit changes in cycle length to not exceed a user-specified value.	

Con Ops Reference Number	Concept of Operations Sample Statements	System Requirements (Tailor as required - See Guidance)	Guidance Section
		2.2.0-5.0-4.0-1 (Sequence-based only) The ASCT shall increase the limit for the following XX cycles based on a change in conditions.	
		2.2.0-5.0-4.0.1.0-2 (Sequence-based only) The increased limit shall be user-defined.	
		2.2.0-5.0-4.0.1.0-1 (Sequence-based only) The change in conditions shall be defined by XX successive adaptive increases in cycle length at the maximum rate.	
		2.1.1.0-10 The ASCT shall determine the order of phases at a user-specified intersection. (The calculation will be based on the optimization function.)	
		2.1.3.0-5 The ASCT shall meter traffic into user-specified bottlenecks by storing queues at user-specified locations.	
		2.1.3.0-6 The ASCT shall store queues at user-specified locations.	

Con Ops Reference Number	Concept of Operations Sample Statements	System Requirements (Tailor as required - See Guidance)	Guidance Section
4.1.0-1.0-5	• Manage the locations of queues within the network Note to user when selecting these requirements: Select from requirements in the 2.2 group when sequence-based systems are allowed (sequence-based systems explicitly calculate cycle, offset, and split). Select from requirements in the 2.3 group when non-sequence-based systems are allowed (non-sequence-based systems do not explicitly calculate cycle, offset, and split). (Select requirements from both groups when the vendor is given the choice of supplying one type of adaptive operation or the other.)	2.1.3.0-2 When queues are detected at user-specified locations, the ASCT shall execute user-specified timing plan/operational mode. 2.2.0-3 (Sequence-based only) The ASCT shall calculate phase lengths for all phases at each signal controller to suit the current coordination strategy. 2.1.3.0-1 The ASCT shall detect the presence of queues at pre-configured locations. 2.1.3.0-3 When queues are detected at user-specified locations, the ASCT shall execute user-specified adaptive operation strategy. 2.1.3.0-4 When queues are detected at user-specified locations, the ASCT shall omit a user-specified phase at a user-specified signal controller. 2.1.3.0-5 he ASCT shall meter traffic into user-specified bottlenecks by storing queues at user-specified locations. 2.1.3.0-6 The ASCT shall store queues at user-specified locations. 2.1.3.0-8 When queues are detected at user-specified locations, the ASCT shall limit the cycle length of the group to a user-specified value.	3.4 3.5

Con Ops Reference Number	Concept of Operations Sample Statements	System Requirements (Tailor as required - See Guidance)	Guidance Section
4.1.0.1.0-6	• At an isolated intersection, optimize operation with a minimum of phase failures (based on the optimization objectives).	2.4.0.2 The ASCT shall calculate a cycle length of a single intersection, based on current measured traffic conditions. (The calculation is based on the optimization objectives.)	3.4
			3.5
		2.4.0.3 The ASCT shall calculate optimum phase lengths, based on current measured traffic conditions. (The calculation is based on the optimization objectives.)	
		2.4.0.4 The ASCT shall calculate phase order, based on current measured traffic conditions. (The calculation is based on the optimization objectives.)	
		2.1.1.0-8.0-1 The ASCT shall provide a user-specified maximum value for each phase at each signal controller.	
		2.1.1.0-8.0-1.0-1 The ASCT shall not provide a phase length longer that the maximum value.	
		2.1.1.0-8.0-2 The ASCT shall provide a user-specified minimum value for each phase at each signal controller.	
		2.1.1.0-8.0-2.0-1 The ASCT shall not provide a phase length shorter than the minimum value.	
		2.1.1.0-8 The ASCT shall provide maximum and minimum phase times.	
		2.4.0.3.0-1 The ASCT shall limit the difference between the length of a given phase and the length of the same phase during its next service to a user-specified value.	

Con Ops Reference Number	Concept of Operations Sample Statements	System Requirements (Tailor as required - See Guidance)	Guidance Section
		2.4.0-3.0-2 When queues are detected at user-specified locations, the ASCT shall execute user-specified timing plan/operational mode.	
4.1.0-2	The system operator needs to manage the coordination in small groups of signals to link phase service at some intersections with phase service at adjacent intersections.

Note that phase-based systems do not explicitly calculate cycle, offset and split at all intersections. | 2.5.0-3 (Phase-based only) The ASCT shall calculate the time at which a user-specified phase shall be green at an intersection. | 3.4 |
		2.5.0-2 (Phase-based only) The ASCT shall alter the state of the signal controller for all phases at the user-specified intersection.	3.5
		2.5.0-4 (Phase-based only) When demand is present, the ASCT shall implement a user-specified maximum time between successive displays of each phase at each intersection.	
		2.5.0-5 (Phase-based only) The ASCT shall alter the operation of the non-critical intersections to minimize stopping of traffic released from user-specified phases at the user-specified critical intersection.	
		2.5.0-6 (Phase-based only) The ASCT shall alter the operation of the non-critical intersections to minimize stopping of traffic arriving at user-specified phases at the user-specified critical intersection.	
		2.5.0-7 (Phase-based only) The ASCT shall adjust the state of the signal controller so that vehicles approaching a signal that have been served during a user-specified phase at an upstream signal do not stop.	

Con Ops Reference Number	Concept of Operations Sample Statements	System Requirements (Tailor as required - See Guidance)	Guidance Section
4.1.0-3	The system operator needs to change the operational strategy (for example, from smooth flow to maximizing throughput or managing queues) based on changing traffic conditions.	2.1.1.0-7.0-1 When current measured traffic conditions meet user-specified criteria, the ASCT shall alter the state of the signal controllers, maximizing the throughput of the coordinated route.	3.4
		2.1.1.0-7.0-2 When current measured traffic conditions meet user-specified criteria, the ASCT shall alter the state of signal controllers, preventing queues from exceeding the storage capacity at user-specified locations.	3.5
		2.1.1.0-7.0-3 When current measured traffic conditions meet user-specified criteria, the ASCT shall alter the state of signal controllers providing equitable distribution of green times.	
		2.1.1.0-7.0-4 When current measured traffic conditions meet user-defined criteria, the ASCT shall alter the state of signal controllers providing two-way progression on a coordinated route.	
		2.1.1.0-7 The ASCT shall alter the adaptive operation to achieve required objectives in user-specified conditions. (The required objectives are specified in Needs Statement 4.1.0-1. Responding to this requirement demonstrates how the proposed system allows the user to define the conditions at which the objectives shift and their associated requirements are fulfilled.) (The alteration may be made by adjusting parameters or by directly controlling the state of signal controllers.)	

Con Ops Reference Number	Concept of Operations Sample Statements	System Requirements (Tailor as required - See Guidance)	Guidance Section
4.1.0-4	The system operator needs to detect repeated phase failures and control signal timing to prevent phase failures building up queues. The operator in this case is trying to prevent a routine queue from forming where it will block another movement in the cycle unnecessarily. For example, the operator may need to prevent a queue resulting from the trailing end of the through green from blocking the storage needed by an entering side-street left turn in the subsequent phase. An overall queue management strategy, particularly when congestion is present, is covered under 4.1.0-1.0-5.	2.1.3.0-2 When queues are detected at user-specified locations, the ASCT shall execute user-specified timing plan/operational mode. 2.2.0-3 (Sequence-based only) The ASCT shall calculate phase lengths for all phases at each signal controller to suit the current coordination strategy. 2.1.3.0-1 The ASCT shall detect the presence of queues at pre-configured locations. 2.1.1.0-9 The ASCT shall detect repeated phases that do not serve all waiting vehicles. (These phase failures may be inferred, such as by detecting repeated max-out.) 2.1.1.0-9.0-1 The ASCT shall alter operations, to minimize repeated phase failures. 2.1.3.0-3 When queues are detected at user-specified locations, the ASCT shall execute user-specified adaptive operation strategy. 2.1.3.0-4 When queues are detected at user-specified locations, the ASCT shall omit a user-specified phase at a user-specified signal controller.	3.4 3.5
4.1.0-5	The system operator needs to minimize the chance that a queue forms at a specified location.	2.3.0-5 (Non-sequence-based only) The ASCT shall adjust signal timing so that vehicles approaching a signal that have been served during a user-specified phase at an upstream signal do not stop.	3.4 3.5

Con Ops Reference Number	Concept of Operations Sample Statements	System Requirements (Tailor as required - See Guidance)	Guidance Section
	Note to user when selecting these requirements:		
	Select from requirements in the 2.2 group when sequence-based systems are allowed (sequence-based systems explicitly calculate cycle, offset, and split).		
	Select from requirements in the 2.3 group when non-sequence-based systems are allowed (non-sequence-based systems do not explicitly calculate cycle, offset, and split).		
	Select from requirements in the 2.5 group when phase-based systems are allowed (phase-based systems do not explicitly calculate cycle, offset and split at all intersections).	2.5.0.7 (Phase-based only) The ASCT shall adjust the state of the signal controller so that vehicles approaching a signal that have been served during a user-specified phase at an upstream signal do not stop.	
	(Select requirements from two or all three groups when the vendor is given the choice of supplying the type of adaptive operation.)	2.2.0.5.0.5 (Sequence-based only) The ASCT shall adjust offsets to minimize the chance of stopping vehicles approaching a signal that have been served by a user-specified phase at an upstream signal.	
4.1.0-6	The system operator needs to modify the sequence of phases to support the various operational strategies.	7.0-6 The ASCT shall provide a minimum of XX different user-defined phase sequences for each signal.	3.4 3.5
		7.0-6.0-1 Each permissible phase sequence shall be user-assignable to any signal timing plan.	
		7.0-6.0-2 Each permissible phase sequence shall be executable by a time of day schedule.	
		7.0-6.0-3 Each permissible phase sequence shall be executable based on measured traffic conditions	
		7.0-7 The ASCT shall not prevent a phase/overlap output by time-of-day.	

Con Ops Reference Number	Concept of Operations Sample Statements	System Requirements (Tailor as required - See Guidance)	Guidance Section
		7.0.8 The ASCT shall not prevent a phase/overlap output based on an external input.	
		7.0.9 The ASCT shall not prevent the following phases to be designated as coordinated phases. (User to list all required phases.)	
4.1.0.7	The system operator needs to fix the sequence of phases at any specified location. For example, the operator may need to fix the phase order at a diamond interchange.	2.1.2.0-12 The ASCT shall not alter the order of phases at a user-specified intersection.	3.4 3.5
4.1.0.8	The system operator needs to designate the coordinated route based on traffic conditions and the selected operational strategy	2.1.1.0-11 The ASCT shall provide coordination along a route.	3.4 3.5
		2.1.1.0-11.0-1 The ASCT shall coordinate along a user-defined route.	
		2.1.1.0-11.0-2 The ASCT shall determine the coordinated route based on traffic conditions.	
		2.1.1.0-11.0-3 The ASCT shall determine the coordinated route based on a user-defined schedule.	
		2.1.1.0-11.0-4 The ASCT shall store XX user-defined coordination routes.	
		2.1.1.0-11.0-4.0-1 The ASCT shall implement a stored coordinated route by operator command.	
		2.1.1.0-11.0-4.0-2 The ASCT shall implement a stored coordinated route based on traffic conditions.	

Con Ops Reference Number	Concept of Operations Sample Statements	System Requirements (Tailor as required - See Guidance)	Guidance Section
		2.1.1.0-11.0-4.0-3 The ASCT shall implement a stored coordinated route based on a user-defined schedule.	3.4
4.1.0-9	The system operator needs to set signal timing parameters (such as minimum green, maximum green and extension time) to comply with agency policies.	2.1.1.0-12 The ASCT shall not prevent the use of phase timings in the local controller set by agency policy.	3.5
4.2	**Network characteristics**		4.1
4.2.0-1	The system operator needs to eventually adaptively control up to XXX signals, up to XXX miles from the TMC (or specified location).	1.0-1 The ASCT shall control a minimum of XX signals concurrently	4.1
4.2.0-2	The system operator needs to be able to adaptively control up to XX independent groups of signals	1.0-2 The ASCT shall support groups of signals.	4.1
		1.0-2.0-2 The ASCT shall control a minimum of XX groups of signals.	
		1.0-2.0-4 Each group shall operate independently	
		1.0-2.0-1 The boundaries surrounding signal controllers that operate in a coordinated fashion shall be defined by the user.	
4.2.0-3	The system operator needs to vary the number of signals in an adaptively controlled group to accommodate the prevailing traffic conditions.	1.0-2 The ASCT shall support groups of signals.	4.1
		1.0-2.0-3 The size of a group shall range from 1 to XX signals.	
		1.0-2.0-5.0-1 The boundaries surrounding signal controllers that operate in a coordinated fashion shall be altered by the system according to a time of day schedule. (For example: this may be achieved by assigning signals to different groups or by combining groups.)	
		1.0-2.0-5.0-2 The boundaries surrounding signal controllers that operate in a coordinated fashion shall be altered by the system according to traffic conditions. (For example: this may be achieved by assigning signals to different groups or by combining groups.)	

Con Ops Reference Number	Concept of Operations Sample Statements	System Requirements (Tailor as required - See Guidance)	Guidance Section
4.3	**4.3 Coordination across boundaries**	1.0-2.0-5 The boundaries surrounding signal controllers that operate in a coordinated fashion shall be altered by the ASCT system according to configured parameters. 1.0-2.0-5.0-3 The boundaries surrounding signal controllers that operate in a coordinated fashion shall be altered by the system when commanded by the user.	
4.3.0-1	The system operator needs to adaptively control signals operated by (specify jurisdictions).	3.0-1 The ASCT shall support external interfaces according to the referenced interface control documents and the following detailed requirements. (Insert appropriate requirements that suit your needs. Interface data flows should be documented in your ITS architecture. Interface requirements include: • Information layer protocol • Application layer protocol • Lower layer protocol • Data aggregation • Frequency of storage • Frequency of reporting • Duration of storage)	4.2 4.3 4.2 4.3
4.3.0-2	The system operator needs to send data to another system that would allow the other system to coordinate with the ASCT system.	3.0-1 The ASCT shall support external interfaces according to the referenced interface control documents and the following detailed requirements. (Insert appropriate requirements that suit your needs. Interface data flows should be documented in your ITS architecture. Interface requirements include: • Information layer protocol • Application layer protocol • Lower layer protocol • Data aggregation • Frequency of storage • Frequency of reporting • Duration of storage)	4.2 4.3

Con Ops Reference Number	Concept of Operations Sample Statements	System Requirements (Tailor as required - See Guidance)	Guidance Section
		3.0-1.0-1 The ASCT shall send operational data to XX external system. (Insert appropriate requirements that suit your needs.)	
		3.0-1.0-2 The ASCT shall send control data to the XX external system. (Insert appropriate requirements that suit your needs.)	
		3.0-1.0-4 The ASCT shall send coordination data to the XX external system. (Insert appropriate requirements that suit your needs.)	
4.3.0-3	The system operator needs to adaptively coordinate signals on two crossing routes simultaneously. (Include signals on crossing arterials within the boundaries of the adaptive systems mapped in Chapter 3.)	4.0-1.0-4 The ASCT shall support adaptive coordination on crossing routes.	4.2 4.3
4.3.0-4	The system operator needs to receive data from another system that will allow the ASCT system to coordinate its operation with the adjacent system.	3.0-1 The ASCT shall support external interfaces according to the referenced interface control documents and the following detailed requirements. (Insert appropriate requirements that suit your needs. Interface data flows should be documented in your ITS architecture. Interface requirements include: • Information layer protocol • Application layer protocol • Lower layer protocol • Data aggregation • Frequency of storage • Frequency of reporting • Duration of storage) 4.0.1.0-1 The ASCT shall alter its operation to minimize interruption of traffic entering the system. (This may be achieved via detection, with no direct connection to the other system.)	4.2 4.3

Con Ops Reference Number	Concept of Operations Sample Statements	System Requirements (Tailor as required - See Guidance)	Guidance Section
4.3.0-5	The system operator needs to constrain the adaptive system to operate a cycle length compatible with the crossing arterial.	4.0-1 The ASCT shall conform its operation to an external system's operation.	
		4.0.1.0-3 The ASCT shall alter its operation based on data received from another system.	
		4.0.1.0-2 The ASCT shall operate a fixed cycle length to match the cycle length of an adjacent system.	4.2 4.3
4.3.0-6	The system operator needs to detect traffic approaching from a neighboring system and coordinate the ASCT operation with the adjacent system.	4.0.1.0-1 The ASCT shall alter its operation to minimize interruption of traffic entering the system. (This may be achieved via detection, with no direct connection to the other system.)	4.2 4.3
		4.0-1 The ASCT shall conform its operation to an external system's operation.	
4.4	Security		
4.4.0-1	The system operator needs to have a security management and administrative system that allows access and operational privileges to be assigned, monitored and controlled by an administrator, and conform to the agency's access and network infrastructure security policies.	5.0-1 The ASCT shall be implemented with a security policy that addresses the following selected elements:	4.3.4
		5.0.1.0-1 • Local access to the ASCT	4.3.4
		5.0.1.0-2 • Remote access to the ASCT	
		5.0.1.0-3 • System monitoring	
		5.0.1.0-4 • System manual override	
		5.0.1.0-5 • Development	

Con Ops Reference Number	Concept of Operations Sample Statements	System Requirements (Tailor as required - See Guidance)	Guidance Section
		5.0-1.0-6 • Operations	4.3.4
		5.0-1.0-7 • User login	
		5.0-1.0-8 • User password	
		5.0-1.0-9 • Administration of the system	
		5.0-1.0-10 • Signal controller group access	
		5.0-1.0-11 • Access to classes of equipment	
		5.0-1.0-12 • Access to equipment by jurisdiction	
		5.0-1.0-13 • Output activation	
		5.0-1.0-14 • System parameters	
		5.0-1.0-15 • Report generation	
		5.0-1.0-16 • Configuration	
		5.0-1.0-17 • Security alerts	
		5.0-1.0-18 • Security logging	
		5.0-1.0-19 • Security reporting	
		5.0-1.0-20 • Database	

Con Ops Reference Number	Concept of Operations Sample Statements	System Requirements (Tailor as required - See Guidance)	Guidance Section
		5.0-1.0-20 • Database 5.0-1.0-21 • Signal controller 5.0-3 The ASCT shall comply with the agency's security policy as described in (specify appropriate policy document).	
4.5	Queuing interactions		
4.5.0-1	The system operator needs to detect queues from outside the system and modify the ASCT operation to accommodate the queuing.	2.1.3.0-2 When queues are detected at user-specified locations, the ASCT shall execute user-specified timing plan/operational mode. 2.1.3.0-1 The ASCT shall detect the presence of queues at pre-configured locations. 2.1.3.0-3 When queues are detected at user-specified locations, the ASCT shall execute user-specified adaptive operation strategy.	4.4 4.4
4.5.0-2	The system operator needs to detect queues within the system's boundaries and modify the ASCT operation to accommodate the queuing.	2.1.3.0-2 When queues are detected at user-specified locations, the ASCT shall execute user-specified timing plan/operational mode. 2.1.3.0-1 The ASCT shall detect the presence of queues at pre-configured locations. 2.1.3.0-3 When queues are detected at user-specified locations, the ASCT shall execute user-specified adaptive operation strategy.	4.4

Con Ops Reference Number	Concept of Operations Sample Statements	System Requirements (Tailor as required - See Guidance)	Guidance Section
4.5.0.3	The system operator needs to detect queues propagating outside its boundaries from within the ASCT boundaries, and modify its operation to accommodate the queuing.	2.1.3.0.2 When queues are detected at user-specified locations, the ASCT shall execute user-specified timing plan/operational mode. 2.1.3.0-1 The ASCT shall detect the presence of queues at pre-configured locations. 2.1.3.0.3 When queues are detected at user-specified locations, the ASCT shall execute user-specified adaptive operation strategy.	4.4
4.5.0.4	The system operator needs to store queues in locations where they can be accommodated without adversely affecting adaptive operation	2.1.3.0.2 When queues are detected at user-specified locations, the ASCT shall execute user-specified timing plan/operational mode. 2.1.3.0-1 The ASCT shall detect the presence of queues at pre-configured locations. 2.1.3.0.3 When queues are detected at user-specified locations, the ASCT shall execute user-specified adaptive operation strategy. 2.1.3.0.4 When queues are detected at user-specified locations, the ASCT shall omit a user-specified phase at a user-specified signal controller. 2.1.3.0.5 The ASCT shall meter traffic into user-specified bottlenecks by storing queues at user-specified locations. 2.1.3.0.6 The ASCT shall store queues at user-specified locations. 2.1.3.0.7 The ASCT shall maintain capacity flow through user-specified bottlenecks.	4.4

Con Ops Reference Number	Concept of Operations Sample Statements	System Requirements (Tailor as required - See Guidance)	Guidance Section
4.5.0-5	The system operator needs to prevent queues forming at user-specified locations.	2.1.3.0-2 When queues are detected at user-specified locations, the ASCT shall execute user-specified timing plan/operational mode.	4.4
		2.1.3.0-1 The ASCT shall detect the presence of queues at pre-configured locations.	
		2.1.3.0-3 When queues are detected at user-specified locations, the ASCT shall execute user-specified adaptive operation strategy.	
		2.1.3.0-4 When queues are detected at user-specified locations, the ASCT shall omit a user-specified phase at a user-specified signal controller.	
		2.1.3.0-5 The ASCT shall meter traffic into user-specified bottlenecks by storing queues at user-specified locations.	
		2.1.3.0-6 The ASCT shall store queues at user-specified locations.	
		2.1.3.0-7 The ASCT shall maintain capacity flow through user-specified bottlenecks.	
4.6	Pedestrians		4.5
4.6.0-1	The system operator needs to accommodate infrequent pedestrian operation and then adaptively recover. (This is appropriate for rare pedestrian calls.)	8.0-3 When a pedestrian phase is called, the ASCT shall accommodate pedestrian crossing times then resume adaptive operation.	4.5
4.6.0-2	The system operator needs to accommodate infrequent pedestrian operation while maintaining adaptive operation. (This is appropriate for pedestrian calls that are common but not so frequent that they drive the operational needs.)	8.0-2 When a pedestrian phase is called, the ASCT shall accommodate pedestrian crossing times during adaptive operations.	4.5

Con Ops Reference Number	Concept of Operations Sample Statements	System Requirements (Tailor as required - See Guidance)	Guidance Section
4.6.0.3	The system operator needs to incorporate frequent pedestrian operation into routine adaptive operation. (This is appropriate when pedestrians are frequent enough that they must be assumed to be present every cycle or nearly every cycle.)	8.0-2 When a pedestrian phase is called, the ASCT shall accommodate pedestrian crossing times during adaptive operations.	4.5
		8.0-5 The ASCT shall execute pedestrian recall on user-defined phases in accordance with a time of day schedule.	
		8.0-7 When specified by the user, the ASCT shall execute pedestrian recall on pedestrian phase adjacent to coordinated phases.	
		8.0-8 When the pedestrian phases are on recall, the ASCT shall accommodate pedestrian timing during adaptive operation.	
4.6.0.4	The system operator needs to accommodate the following custom pedestrian features. (Describe custom features in this need and then create appropriate requirements.)		4.5
4.6.0.5	The system operator needs to accommodate early start of walk and exclusive pedestrian phases.	8.0-1 When a pedestrian phase is called, the ASCT shall execute pedestrian phases up to XX seconds before the vehicle green of the related vehicle phase.	4.5
		8.0-4 The ASCT shall execute user-specified exclusive pedestrian phases during adaptive operation.	
4.7 Non-adaptive situations			
4.7.0-1	The system operator needs to detect traffic conditions during which adaptive control is not the preferred operation, and implement some pre-defined operation while that condition is present.	2.1.1.0-1 The ASCT shall operate non-adaptively during the presence of a defined condition.	4.6
4.7.0-2	The system operator needs to schedule pre-determined operation by time of day.	2.1.1.0-5 The ASCT shall operate non-adaptively in accordance with a user-defined time-of-day schedule.	4.6

Con Ops Reference Number	Concept of Operations Sample Statements	System Requirements (Tailor as required - See Guidance)	Guidance Section
4.7.0.3	The system operator needs to over-ride adaptive operation.	2.1.1.0-3 The ASCT shall operate non-adaptively when a user manually commands the ASCT to cease adaptively controlling a group of signals.	4.6
		2.1.1.0-4 The ASCT shall operate non-adaptively when a user manually commands the ASCT to cease adaptive operation.	
		2.1.1.0-5 The ASCT shall operate non-adaptively in accordance with a user-defined time-of-day schedule	
4.8	System responsiveness		4.7
4.8.0-1	The system operator needs to modify the ASCT operation to closely follow changes in traffic conditions.	2.6.0-1 The ASCT shall limit the change in consecutive cycle lengths to be less than a user-specified value.	4.7
		2.6.0-2 The ASCT shall limit the change in phase times between consecutive cycles to be less than a user-specified value. (This does not apply to early gap-out or actuated phase skipping.)	
		2.6.0-3 The ASCT shall limit the changes in the direction of primary coordination to a user-specified frequency.	
4.8.0-2	The system operator needs to constrain the selection of cycle lengths to those that provide acceptable operations, such as when resonant progression solutions are desired.	2.6.0-4 When a large change in traffic demand is detected, the ASCT shall respond more quickly than normal operation, subject to user-specified limits. (DEFINE "MORE QUICKLY")	4.7
4.9	Complex coordination and controller features		4.8

Con Ops Reference Number	Concept of Operations Sample Statements	System Requirements (Tailor as required - See Guidance)	Guidance Section
4.9.0-1	The system operator needs to implement the following advanced controller features while maintaining adaptive operation:		4.8
4.9.0-1.0-1	• Service a phase more than once per cycle	7.0-1 When specified by the user, the ASCT shall serve a vehicle phase more than once for each time the coordinated phase is served.	4.8
4.9.0-1.0-2	• Operate at least XX overlap phases	7.0-2 The ASCT shall provide a minimum of XX phase overlaps.	4.8
4.9.0-1.0-3	• Operate four rings, 16 phases and up to three phases per ring (Edit to suit your needs)	7.0-3 The ASCT shall accommodate a minimum of XX phases at each signal	4.8
		7.0-4 The ASCT shall accommodate a minimum of XX rings at each signal.	
		7.0-5 The ASCT shall accommodate a minimum of XX phases per ring	
4.9.0-1.0-4	• Permit different phase sequences under different traffic conditions	7.0-6 The ASCT shall provide a minimum of XX different user-defined phase sequences for each signal.	4.8
		7.0-6.0-1 Each permissible phase sequence shall be user-assignable to any signal timing plan.	
		7.0-6.0-2 Each permissible phase sequence shall be executable by a time of day schedule.	
		7.0-6.0-3 Each permissible phase sequence shall be executable based on measured traffic conditions	

Con Ops Reference Number	Concept of Operations Sample Statements	System Requirements (Tailor as required - See Guidance)	Guidance Section
4.9.0.1.0-5	• Allow one or more phases to be omitted (disabled) under certain traffic conditions or signal states.	2.1.2.0-6 The ASCT shall omit a user-specified phase when the cycle length is below a user-specified value.	4.8
		2.1.2.0-9 The ASCT shall omit a user-specified phase according to a time of day schedule	
		2.1.2.0-7 The ASCT shall omit a user-specified phase based on measured traffic conditions.	
		2.1.2.0-8 The ASCT shall omit a user-specified phase based on the state of a user-specified external input.	
4.9.0.1.0-6	• Prevent one or more phases being skipped under certain traffic conditions or signal states.	2.1.2.0-5 The ASCT shall prevent skipping a user-specified phase according to a time of day schedule.	4.8
		2.1.2.0-3 The ASCT shall prevent skipping a user-specified phase when the user-specified phase sequence is operating.	
		2.1.2.0-4 The ASCT shall prevent skipping a user-specified phase based on the state of a user-specified external input.	
4.9.0.1.0-7	• Allow detector logic at an intersection to be varied depending on local signal states	7.0-15 The ASCT shall operate adaptively with the following detector logic. (DESCRIBE THE CUSTOM LOGIC)	4.8
4.9.0.1.0-8	• Accommodate the following custom features used by this agency (describe the features)	7.0-14 (Describe requirements to suit other custom controller features that must be accommodated.)	4.8
4.9.0.1.0-9	• Allow any phase to be designated as the coordinated phase	7.0-9 The ASCT shall not prevent the following phases to be designated as coordinated phases. (User to list all required phases.)	4.8

Con Ops Reference Number	Concept of Operations Sample Statements	System Requirements (Tailor as required - See Guidance)	Guidance Section
4.9.0.1.0-10	• Allow the operator to specify which phase receives unused time from a preceding phase	2.1.2.0-10 The ASCT shall assign unused time from a preceding phase that terminates early to a user-specified phase as follows: • Next phase • Next coordinated phase • User-specified phase 2.1.2.0-11 The ASCT shall assign unused time from a preceding phase that is skipped to a user-specified phase as follows: • Previous phase • Next phase • Next coordinated phase • User-specified phase	4.8
4.9.0.1.0-11	• Allow the controller to respond independently to individual lanes of an approach. This may be implemented in the signal controller using XX extension/passage timers, which may be assignable to each vehicle detector input channel. This may allow the adaptive operation to be based on data from a specific detector, or by excluding specific detectors.	7.0-12 The ASCT shall not prevent the local signal controller from performing actuated phase control using XX extension/passage timers as assigned to user-specified vehicle detector input channels in the local controller. 9.0-1 The ASCT shall set a specific state for each special function output based on the occupancy on a user-specified detector. 7.0-12.0-1 The ASCT shall operate adaptively using user-specified detector channels.	4.8
4.9.0.1.0-12	• Allow the coordinated phase to terminate early under prescribed traffic conditions	7.0-10 The ASCT shall have the option for a coordinated phase to be released early based on a user-definable point in the phase or cycle. (User select phase or cycle.)	4.8
4.9.0.1.0-13	• Allow flexible timing of non-coordinated phases (such as late start of a phase) while maintaining coordination	8.0-6 The ASCT shall begin a non-coordinated phase later than its normal starting point within the cycle when all of the following conditions exist:	4.8

Con Ops Reference Number	Concept of Operations Sample Statements	System Requirements (Tailor as required - See Guidance)	Guidance Section
		• The user enables this feature • Sufficient time in the cycle remains to serve the minimum green times for the phase and the subsequent non-coordinated phases before the beginning of the coordinated phase • The phase is called after its normal start time • The associated pedestrian phase is not called	
4.9.0.1.0-14	• Protected/permissive phasing and alternate left turn phase sequences.	2.1.2.0-1 The ASCT shall not prevent protected/permissive left turn phase operation. 2.1.2.0-2 The ASCT shall not prevent the protected left turn phase to lead or lag the opposing through phase based upon user-specified conditions.	4.8
4.9.0.1.0-15	• Use flashing yellow arrow to control permissive left turns and right turns.	7.0-11 The ASCT shall not prevent the controller from displaying flashing yellow arrow left turn or right turn. (SELECT AS APPLICABLE)	4.8
4.9.0.1.0-16	• Service side streets and pedestrian phases at minor locations more often than at adjacent signals when this can be done without compromising the quality of the coordination. (E.g., double-cycle mid-block pedestrian crossing signals.)	7.0-13 When adaptive operation is used in conjunction with normal coordination, the ASCT shall not prevent a controller serving a cycle length different from the cycles used at adjacent intersections.	4.8
4.9.0.1.0-17	• Use negative pedestrian phasing to prevent an overlap conflicting with a pedestrian walk/don't walk	8.0-9 The ASCT shall not inhibit negative vehicle and pedestrian phase timing.	4.8
4.10	4.10 Monitoring and control		4.9
4.10.0-1	The system operator needs to monitor and control all required features of adaptive operation from the following locations: (Edit and select as appropriate to suit your situation.)	5.0.2 The ASCT shall provide monitoring and control access at the following locations:	4.9
4.10.0.1.0-1	• Agency TMC	5.0.2.0-1 • Agency TMC	4.9

Con Ops Reference Number	Concept of Operations Sample Statements	System Requirements (Tailor as required - See Guidance)	Guidance Section
4.10.0-1.0-2	• Maintenance facility	5.0-2.0-2 • Maintenance facility	4.9
4.10.0-1.0-3	• Workstations on agency LAN or WAN located at (specify)	5.0-2.0-3 • Agency LAN or WAN	4.9
4.10.0-1.0-4	• Other agency's TMC (specify)	5.0-2.0-4 • Other agency TMC	4.9
4.10.0-1.0-5	• Local controller cabinets	5.0-2.0-5 • Local controller cabinets	4.9
4.10.0-1.0-6	• Maintenance vehicles	5.0-2.0-6 • Maintenance vehicles	4.9
4.10.0-1.0-7	• Remote locations (specify)	5.0-2.0-7 • Remote locations via internet	4.9
4.10.0-2	The operator needs to access to the database management, monitoring and reporting features and functions of the signal controllers and any related signal management system from the access points defined for those system components.	5.0-4 The ASCT shall not prevent access to the local signal controller database, monitoring or reporting functions by any installed signal management system.	4.9
4.11	4.11 Performance reporting		4.10
4.11.0-1	The agency needs the (specify external decision support system) to be able to monitor the ASCT system automatically.	3.0-1.0-3 The ASCT shall send monitoring data to the XX external system. (Insert appropriate requirements that suit your needs.)	4.10

Con Ops Reference Number	Concept of Operations Sample Statements	System Requirements (Tailor as required - See Guidance)	Guidance Section
4.11.0-2	The system operator needs to store and report data used to calculate signal timing and have the data available for subsequent analysis.	6.0-4 The ASCT shall store results of all signal timing parameter calculations for a minimum of XX days. 6.0-5 The ASCT shall store the following measured data in the form used as input to the adaptive algorithm for a minimum of XX days: (edit as appropriate) • Volume • Occupancy • Queue length • Phase utilization • Arrivals in green • Green band efficiency 6.0-12 The ASCT shall store the following data in XX minute increments: (edit as appropriate) • Volume • Occupancy • Queue length 18.0-1 The ASCT shall report measures of current traffic conditions on which it bases signal state alterations. 18.0-2 The ASCT shall report all intermediate calculated values that are affected by calibration parameters. 18.0-3 The ASCT shall maintain a log of all signal state alterations directed by the ASCT.	4.10
4.11.0-3	The system operator needs to store and report data that can be used to measure traffic performance under adaptive control.	6.0-4 The ASCT shall store results of all signal timing parameter calculations for a minimum of XX days.	4.10

Con Ops Reference Number	Concept of Operations Sample Statements	System Requirements (Tailor as required - See Guidance)	Guidance Section
4.11.0-4	The system operator needs to store all operational data and signal timing parameters calculated by the adaptive system, and export selected data to (specify appropriate external system).	**6.0-5** The ASCT shall store the following measured data in the form used as input to the adaptive algorithm for a minimum of XX days: (edit as appropriate) • Volume • Occupancy • Queue length • Phase utilization • Arrivals in green • Green band efficiency **6.0-12** The ASCT shall store the following data in XX minute increments: (edit as apporpriate) • Volume • Occupancy • Queue length **6.0-2** The ASCT shall export its systems log in the following formats: (edit as appropriate) • MS Excel • Text • CVS • Open source SQL database **6.0-3** The ASCT shall store the event log for a minimum of XX days **6.0-6** The ASCT system shall archive all data automatically after a user-specified period not less than XX days.	4.10

Con Ops Reference Number	Concept of Operations Sample Statements	System Requirements (Tailor as required - See Guidance)	Guidance Section
		6.0.7 The ASCT shall provide data storage for a system size of XX signal controllers. The data to be stored shall include the following: (edit as appropriate) • Controller state data • Reports • Log data • Security data • ASCT parameters • Detector status data 6.0-10 The ASCT shall store data logs in a standard database (specify as appropriate).	
4.11.0-5	The system operator needs to report performance data in real time to (specify external system).	3.0-1 The ASCT shall support external interfaces according to the referenced interface control documents and the following detailed requirements. (Insert appropriate requirements that suit your needs. Interface data flows should be documented in your ITS architecture. Interface requirements include: • Information layer protocol • Application layer protocol • Lower layer protocol • Data aggregation • Frequency of storage • Frequency of reporting • Duration of storage) 3.0-1.0-1 The ASCT shall send operational data to XX external system. (Insert appropriate requirements that suit your needs.) 3.0-1.0-5 The ASCT shall send performance data to the XX external system. (Insert appropriate requirements that suit your needs.)	4.10

Con Ops Reference Number	Concept of Operations Sample Statements	System Requirements (Tailor as required - See Guidance)	Guidance Section
4.11.0-6	The system operator needs to be able to report the exact state of signal timing and input data for a specified period, to allow historical analysis of the system operation.	6.0-1 The ASCT shall log the following events: (edit as appropriate) 6.0-1.0-1 Time-stamped vehicle phase calls 6.0-1.0-2 Time-stamped pedestrian phase calls 6.0-1.0-3 Time-stamped emergency vehicle preemption calls 6.0-1.0-4 Time-stamped transit priority calls 6.0-1.0-5 Time-stamped railroad preemption calls 6.0-1.0-6 Time-stamped start and end of each phase 6.0-1.0-7 Time-stamped controller interval changes 6.0-1.0-8 Time-stamped start and end of each transition to a new timing plan	4.10
4.11.0-7	Have the ability to generate historic and real-time reports that effectively support operation, maintenance and reporting of system performance and traffic conditions.	6.0-5 The ASCT shall store the following measured data in the form used as input to the adaptive algorithm for a minimum of XX days: (edit as appropriate) • Volume • Occupancy • Queue Length • Phase Utilization • Arrivals in Green • Green Band Efficiency	4.10

Con Ops Reference Number	Concept of Operations Sample Statements	System Requirements (Tailor as required - See Guidance)	Guidance Section
		6.0-8 The ASCT shall calculate and report relative data quality including: • The extent data is affected by detector faults • Other applicable items	
		6.0-9 The ASCT shall report comparisons of logged data when requested by the user: • Day to day, • Hour to hour • Hour of day to hour of day • Hour of week to hour of week • Day of week to day week • Day of year to day of year	
		6.0-11 The ASCT shall report stored data in a form suitable to provide explanations of system behavior to public and politicians and to troubleshoot the system.	
		18.0.3 The ASCT shall maintain a log of all signal state alterations directed by the ASCT.	
		18.0.3.0-4 The ASCT shall maintain the records in this ASCT log for XX period.	
		18.0.3.0-5 The ASCT shall archive the ASCT log in the following manner: (Specify format, frequency, etc., to suit your needs.)	
		18.0.3.0-1 The ASCT log shall include all events directed by the external inputs.	

Con Ops Reference Number	Concept of Operations Sample Statements	System Requirements (Tailor as required - See Guidance)	Guidance Section
4.12	4.12 Failure notification		
4.12.0-1	The system operator needs to immediately notify maintenance and operations staff of alarms and alerts.	13.1.0-3 In the event of a detector failure, the ASCT shall issue an alarm to user-specified recipients. (This requirement may be fulfilled by sending the alarm to a designated list of recipients by a designated means, or by using an external maintenance management system.	4.11
		13.2-2 In the event of communications failure, the ASCT shall issue an alarm to user-specified recipients. (This requirement may be fulfilled by sending the alarm to a designated list of recipients by a designated means, or by using an external maintenance management system.	
		13.3-2 In the event of adaptive processor failure, the ASCT shall issue an alarm to user-specified recipients. (This requirement may be fulfilled by sending the alarm to a designated list of recipients by a designated means, or by using an external maintenance management system.	
		13.2-3 The ASCT shall issue an alarm within XX minutes of detection of a failure.	
4.12.0-2	The system operator needs to immediately and automatically pass alarms and alerts to the (specify external system).	13.1.0-3 In the event of a detector failure, the ASCT shall issue an alarm to user-specified recipients. (This requirement may be fulfilled by sending the alarm to a designated list of recipients by a designated means, or by using an external maintenance management system.	4.11

Additional rows above 4.12:

| | | 18.0.3.0-2 The ASCT log shall include all external output state changes. | |
| | | 18.0.3.0-3 The ASCT log shall include all actual parameter values that are subject to user-specified values. | 4.11 |

Con Ops Reference Number	Concept of Operations Sample Statements	System Requirements (Tailor as required - See Guidance)	Guidance Section
		13.2-2 In the event of communications failure, the ASCT shall issue an alarm to user-specified recipients. (This requirement may be fulfilled by sending the alarm to a designated list of recipients by a designated means, or by using an external maintenance management system.	
		13.3-2 In the event of adaptive processor failure, the ASCT shall issue an alarm to user-specified recipients. (This requirement may be fulfilled by sending the alarm to a designated list of recipients by a designated means, or by using an external maintenance management system.	
		13.2-3 The ASCT shall issue an alarm within XX minutes of detection of a failure.	
4.12.0-3	The system operator needs to maintain a complete log of alarms and failure events.	13.1.0-4 In the event of a failure, the ASCT shall log details of the failure in a permanent log.	4.11
		13.1.0-5 The permanent failure log shall be searchable, archivable and exportable.	
		13.2-4 In the event of a communications failure, the ASCT shall log details of the failure in a permanent log.	
		13.2-5 The permanent failure log shall be searchable, archivable and exportable.	
4.13	4.13 Preemption and priority		4.12
4.13.0-1	The system operator needs to accommodate railroad and light rail preemption (explain further)	11.0-1 The ASCT shall maintain adaptive operation at non-preempted intersections during railroad preemption.	
		11.0-4 The ASCT shall resume adaptive control of signal controllers when preemptions are released.	

Con Ops Reference Number	Concept of Operations Sample Statements	System Requirements (Tailor as required - See Guidance)	Guidance Section
		11.0.5 The ASCT shall execute user-specified actions at non-preempted signal controllers during preemption. (E.g., inhibit a phase, activate a sign, display a message on a DMS)	
		11.0.6 The ASCT shall operate normally at non-preempted signal controllers when special functions are engaged by a preemption event. (Examples of such special functions are a phase omit, a phase maximum recall or a fire route.)	
		11.0.7 The ASCT shall release user-specified signal controllers to local control when one signal in a group is preempted.	
		11.0.8 The ASCT shall not prevent the local signal controller from operating in normally detected limited-service actuated mode during preemption.	
		11.0.3 The ASCT shall maintain adaptive operation at non-preempted intersections during Light Rail Transit preemption.	
4.13.0-2	The system operator needs to accommodate emergency vehicle preemption (explain further)	11.0.4 The ASCT shall resume adaptive control of signal controllers when preemptions are released.	4.12
		11.0.5 The ASCT shall execute user-specified actions at non-preempted signal controllers during preemption. (E.g., inhibit a phase, activate a sign, display a message on a DMS)	
		11.0.6 The ASCT shall operate normally at non-preempted signal controllers when special functions are engaged by a preemption event. (Examples of such special functions are a phase omit, a phase maximum recall or a fire route.)	

Con Ops Reference Number	Concept of Operations Sample Statements	System Requirements (Tailor as required - See Guidance)	Guidance Section
		11.0.7 The ASCT shall release user-specified signal controllers to local control when one signal in a group is preempted.	
		11.0.8 The ASCT shall not prevent the local signal controller from operating in normally detected limited-service actuated mode during preemption.	
		11.0.2 The ASCT shall maintain adaptive operation at non-preempted intersections during emergency vehicle preemption.	
4.13.0-3	The system operator needs to accommodate bus and light rail transit signal priority (explain further)	12.0-1 The ASCT shall continue adaptive operations of a group when one of its signal controllers has a transit priority call.	4.12
		12.0-2 The ASCT shall advance the start of a user-specified green phase in response to a transit priority call.	
		12.0-3 The ASCT shall delay the end of a green phase, in response to a priority call.	
		12.0-4 The ASCT shall permit at least XX exclusive transit phases.	
		12.0-5 The ASCT shall control vehicle phases independently of the following:	
		12.0-6 The ASCT shall interface with external bus transit priority system in the following fashion...... (explain the external system and refer to other interfaces as appropriate)	
		12.02.0-1 The advance of start of green phase shall be user-defined.	
		12.02.0-2 Adaptive operations shall continue during the advance of the start of green phase.	

Con Ops Reference Number	Concept of Operations Sample Statements	System Requirements (Tailor as required - See Guidance)	Guidance Section
		12.0.3.0-1 The delay of end of green phase shall be user-defined.	
		12.0.3.0-2 Adaptive operations shall continue during the delay of the end of green phase.	
		12.0.4.0-1 Adaptive operations shall continue when there is an exclusive transit phase call.	
		12.0.5.0-1 • LRT only phases	
		12.0.5.0-2 • Bus only phases	
		12.0.8 The ASCT shall accept a transit priority call from: • A signal controller/transit vehicle detector; • An external system.	
		12.0.7 The ASCT shall interface with external light rail transit priority system in the following fashion...... (explain the external system and refer to other interfaces as appropriate)	
4.13.0-4	The system operator needs to accommodate light rail priority (explain further)		4.12
4.14	4.14 Failure and fallback		4.13
4.14.0-1	The system operator needs to fall back to TOD or isolated free operation, as specified by the operator, without causing disruption to traffic flow, in the event of equipment, communications and software failure.	13.1.0-2 The ASCT shall use the following alternate data sources for operations in the absence of the real-time data from a detector:	4.13

Model Systems Engineering Documents for Adaptive Signal Control Technology (ASCT) Systems

Con Ops Reference Number	Concept of Operations Sample Statements	System Requirements (Tailor as required - See Guidance)	Guidance Section
		13.1.0-2.0-3 The ASCT shall switch to the alternate source in real time without operator intervention.	4.13
		13.1.0-1 The ASCT shall take user-specified action in the absence of valid detector data from XX vehicle detectors within a group. (SELECT THE APPROPRIATE ACTION.)	
		13.1.0-1.0-1 The ASCT shall release control to central system control.	
		13.2-1 The ASCT shall execute user-specified actions when communications to one or more signal controllers fails within a group. (SELECT THE APPROPRIATE ACTION)	
		13.2-1.0-1 In the event of loss of communication to a user-specified signal controller, the ASCT shall release control of all signal controllers within a user-specified group to local control.	
		13.3-1 The ASCT shall execute user-specified actions when adaptive control fails:	
		13.3-1.0-1 The ASCT shall release control to central system control.	
		2.1.1.0-2 The ASCT shall operate non-adaptively when adaptive control equipment fails.	
		2.1.1.0-2.0-1 The ASCT shall operate non-adaptively when a user-specified detector fails.	

Con Ops Reference Number	Concept of Operations Sample Statements	System Requirements (Tailor as required - See Guidance)	Guidance Section
		2.1.1.0-2.0.2 The ASCT shall operate non-adaptively when the number of failed detectors connected to a signal controller exceeds a user-defined value.	
		2.1.1.0-2.0.3 The ASCT shall operate non-adaptively when the number of failed detectors in a group exceeds a user-defined value.	
		2.1.1.0-2.0.4 The ASCT shall operate non-adaptively when a user-defined communications link fails.	
		13.1.0.2.0-1 • Data from a user-specified alternate detector.	
		13.1.0.2.0-2 • Stored historical data from the failed detector.	
		13.1.0.1.0-2 The ASCT shall release control to local operations to operate under its own time-of-day schedule.	
		13.2-1.0-2 The ASCT shall switch to the alternate operation in real time without operator intervention.	
		13.3-1.0-2 The ASCT shall release control to local operations to operate under its own time-of-day schedule.	
		13.3-4 During adaptive processor failure, the ASCT shall provide all local detector inputs to the local controller.	
4.15	4.15 Constraints		4.14
4.15.0-1	The system operator is constrained to use the following equipment:		4.14

Con Ops Reference Number	Concept of Operations Sample Statements	System Requirements (Tailor as required - See Guidance)	Guidance Section
4.15.0-1.0-1	• Controller type (list acceptable equipment)	14.0-3 The ASCT shall fully satisfy all requirements when connected with XX controllers (specify controller types).	4.14
4.15.0-1.0-2	• Detector type (list acceptable equipment)	14.0-2 The ASCT shall fully satisfy all requirements when connected with detectors from manufacturer XX (specify required detector types).	4.14
4.15.0-1.0-3	• Communication system (list acceptable equipment)		4.14
4.15.0-1.0-4	• Cabinet type and size (list acceptable equipment)		4.14
4.15.0-1.0-5	• Signal management system (list acceptable systems)		4.14
4.15.0-2	The system operator needs to use equipment and software acceptable under current agency IT policies and procedures.	14.0-1 The vendor's adaptive software shall be fully operational within the following platform: (edit as appropriate) • Windows-PC, • Linux, • Mac-OS, • Unix.	4.14
4.15.0-3	Not used		
4.15.0-4	Not used		
4.16	4.16 Training and support		
4.16.0-1	The agency needs all staff involved in operation and maintenance to receive appropriate training.	15.0-1.0-1 The vendor shall provide training on the operations of the adaptive system. 15.0-1.0-9 The vendor shall provide a minimum of XX hours training to a minimum of XX staff. (specify how much training will be required) 15.0-1 The vendor shall provide the following training. (Edit as appropriate.) 15.0-1.0-2 The vendor shall provide training on troubleshooting the system.	

Con Ops Reference Number	Concept of Operations Sample Statements	System Requirements (Tailor as required - See Guidance)	Guidance Section
		15.0.1.0-3 The vendor shall provide training on preventive maintenance and repair of equipment.	
		15.0.1.0-4 The vendor shall provide training on system configuration.	
		15.0.1.0-5 The vendor shall provide training on administration of the system.	
		15.0.1.0-6 The vendor shall provide training on system calibration.	
		15.0.1.0-7 The vendor's training delivery shall include: printed course materials and references, electronic copies of presentations and references.	
		15.0.1.0-8 The vendor's training shall be delivered at (specify locations for training).	
		15.0.1.0-10 The vendor shall provide a minimum of XX training sessions (specify how many sessions over what period).	
4.16.0-2	The agency needs the system to fulfill all requirements for the life of the system. The agency therefore needs the system to be maintained to repair faults that are not defects in materials and workmanship.	16.0-1 The Maintenance Vendor shall provide maintenance according to a separate maintenance contract. That contract should identify repairs necessary to preserve requirements fulfillment, responsiveness in effecting those repairs, and all requirements on the maintenance provider while performing the repairs.	
4.16.0-3	The agency needs the system to fulfill all requirements for the life of the system. The agency therefore needs the system to remain free of defects in materials and workmanship that result in requirements no longer being fulfilled.	16.0-3 The Vendor shall warrant the system to be free of defects in materials and workmanship for a period of XX years. Warranty is defined as correcting defects in materials and workmanship (subject to other language included in the purchase documents). Defect is defined as any circumstance in which the material does not perform according to its specification.	

Con Ops Reference Number	Concept of Operations Sample Statements	System Requirements (Tailor as required - See Guidance)	Guidance Section
4.16.0-4	The agency needs the system to fulfill all requirements for the life of the system. The agency therefore needs support to keep software and software environment updated as necessary to prevent requirements no longer being fulfilled.	16.0-2 The Vendor shall provide routine updates to the software and software environment necessary to preserve the fulfillment of requirements for a period of XX years. Preservation of requirements fulfillment especially includes all IT management requirements as previously identified.	
4.17	**External interfaces**		
4.17.0-1	The system operator needs to be able to turn on signs that control traffic or provide driver information when specific traffic conditions occur, when needed to support the adaptive operation, when congestion is detected at critical locations or according to a time-of-day schedule	17.0-1 The ASCT shall set the state of external input/output states according to a time-of-day schedule.	
		17.0-2 The ASCT output states shall be settable according to a time-of-day schedule	
		9.0-1 The ASCT shall set a specific state for each special function output based on the occupancy on a user-specified detector.	
		9.0-2 The ASCT shall set a specific state for each special function output based on the current cycle length.	
		9.0-3 The ASCT shall set a specific state for each special function output based on a time-of-day schedule.	
4.17.0-2	The system operator needs to react to commands issued by (specify an external control or decision support system, such as an ICM system or another signal system).	3.0-1 The ASCT shall support external interfaces according to the referenced interface control documents and the following detailed requirements. (Insert appropriate requirements that suit your needs. Interface data flows should be documented in your ITS architecture. Interface requirements include: • Information layer protocol • Application layer protocol • Lower layer protocol • Data aggregation • Frequency of storage • Frequency of reporting • Duration of storage)	

Con Ops Reference Number	Concept of Operations Sample Statements	System Requirements (Tailor as required - See Guidance)	Guidance Section
		7.0-8 The ASCT shall not prevent a phase/overlap output based on an external input.	
		2.1.1.0-6 The ASCT shall operate non-adaptively when commanded by an external system process.	
		4.0-1 The ASCT shall conform its operation to an external system's operation.	
		2.1.2.0-4 The ASCT shall prevent skipping a user-specified phase based on the state of a user-specified external input.	
		2.1.2.0-8 The ASCT shall omit a user-specified phase based on the state of a user-specified external input.	
		3.0-1.0-6 The ASCT shall receive commands from the XX external system.	
		3.0-1.0-7 The ASCT shall implement the following commands from the XX external system when commanded: [Edit as appropriate for your situation) • Specified cycle length • Specified direction of progression • Specified adaptive strategy	
4.18	4.18 Maintenance		
4.18.0-1	Each maintaining agency needs all applicable equipment to be readily accessible.		

Con Ops Reference Number	Concept of Operations Sample Statements
5	**Chapter 5: Envisioned Adaptive System Overview**
5.1	**5.1 Size and grouping**
5.1.0-1	The agency has plans to adaptively control a total of XX intersections.
5.1.0-2	The system will control intersections in groups that are defined by the operator.
5.1.0-3	A group of intersections may be comprised of simply one intersection, or up to the total number of intersections that are sufficiently close to warrant coordination under the prevailing traffic conditions.
5.1.0-4	During some traffic conditions, there may be separate groups of intersections operating with different characteristics (e.g., different cycle lengths, some coordinated some not, offsets in different directions).
5.1.0-5	During periods when traffic conditions are similar or operating characteristics (such as cycle length) are similar, or traffic volumes on the coordinated route are heavier, different groups may be formed or specified by the operator.
5.1.0-6	The group of intersections at XX is XX miles away from the group of intersections at XX. These two groups of intersections will always operate entirely independently.
5.2	**5.2 Operational objectives**
5.2.0-1	The objective of the coordination will be to provide for smooth flow along the arterial road, minimizing the number of stops experienced by vehicles traveling along the road. Where "natural" cycle lengths exist that permit two-way progression, the system will generally operate at one of those cycle lengths unless longer phase lengths are required to accommodate the demand.
5.2.0-2	The objective of the coordination will be to maximize the throughput along the coordinated route. This may involve a tradeoff that increases delay to cross streets and turning movements in order to maximize the green time provided to coordinated traffic flows.
5.2.0-3	The objective of the coordination will be to control traffic in a manner that equitably serves the adjacent land uses. The delays experienced by the traffic entering and leaving the coordinated route will be balanced with the delays and stops experienced by other traffic traveling along the route.
5.2.0-4	The objective of the coordination will be to manage the lengths of queues stored at critical locations within the coordinated group so that long queues do not block upstream intersections or otherwise reduce the capacity available during the green phases. This will involve controlling phase lengths so that the size of platoons entering a downstream block does not exceed the storage length if the platoon will be stopped. It will also involve control of offsets and phase lengths so that queues may be stored in locations where they will not adversely affect capacity of the system.
5.2.0-5	The system, or the operator, will select the appropriate coordination objective, depending on the current traffic conditions. For example, during commuter peaks the primary objective may be to maximize the throughput along the road in the peak direction. Then during the business hours the objective may be to balance delays between traffic associated with the adjacent activity and traffic simply traveling through the system.
5.2.0-6	The operator will be able to define for each group of intersections the appropriate operational objective. For example, near a freeway interchange or in a location with heavy turning movements, the queue management strategy may be specified, while on an arterial with long signal spacing the smooth flow objective may be specified.

Con Ops Reference Number	Concept of Operations Sample Statements
5.2.0-7	During moderate to light traffic conditions, one or more phases may be omitted (e.g., a protected phase if protected/permissive left turns are operated), in order to more efficiently serve other movements, provided it is safe to do so. This may be accomplished through a time of day schedule or based on the measured traffic conditions.
5.2.0-8	Within these operational objectives, the ASCT system will change its operation to accommodate the rise and fall of volumes through the peaks and the changing patterns of flow throughout the day and week. However, there is also a stochastic element to traffic in the short term, with the number of arrivals for a phase varying from cycle to cycle, and pedestrians not being present on all phases in all cycles. It is therefore desirable for the system to have some local tactical control. While vehicle-actuated coordination typically allows phases to run longer or shorter from cycle to cycle to match the actual number of vehicles using the phase, the system will also allow the operator to decide where the unused time will be used. If a phase is to be skipped, the operator can specify that the spare time will be added to the existing phase, the following phase or the next coordinated phase.
5.2.0-9	At an isolated intersection with widely varying traffic patterns and a high degree of saturation during peak times, the system will calculate the optimum cycle length, phase sequence and phase times in real time to match the changing traffic conditions.
5.2.0-10	At a small group of intersections, with the user defining one as being critical, while the adjacent intersections require a lower cycle length or progression must be provided for specific phases to minimize the formation of queues on the approaches to the critical intersection, the phase lengths of the critical intersection will be determined by the system based on the current traffic conditions. The operation of the adjacent intersections will then be set so that platoons departing the critical intersection are progressed through the non-critical intersections, or platoons arriving at the critical intersection do so at a time when they will have little or no delay waiting for the appropriate phase.
5.3	5.3 Fallback operation
5.3.0-1	The system will have a fallback state that allows coordination using a common cycle length for all signals within a coordinated group.
5.3.0-2	The system will have a fallback state that allows individual intersections to operate in a vehicle-actuated, isolated mode in the event of failures of the adaptive processor software or hardware, detectors or communication.
5.3.0-3	The system will have a fallback state that allows one or more intersections to be slaved from a critical intersection in the event of failures of the adaptive processor software or hardware, detectors or communication.
5.4	5.4 Crossing routes and adjacent systems
5.4.0-1	A coordinated group will be able to include more than one coordinated route, such as two crossing arterials. The system will be able to maintain coordination along both roads.
5.4.0-2	The agency needs the adaptive system to maintain coordination with another adjacent system either by sensing arriving traffic or by using constraints on cycle length.
5.4.0-3	The system will accept data from a neighboring system that allows it to stay in coordination with the adjacent system while still operating in adaptive mode.
5.5	5.5 Operator access

Con Ops Reference Number	Concept of Operations Sample Statements
5.5.0-1	Operators, traffic engineering and maintenance staff will be assigned different levels of authority, and access to equipment for which they are authorized, based on their roles and responsibilities. This will allow them to control, view, monitor and analyze the operation of the system as appropriate.
5.5.0-2	The system will a stand-alone system not connected to a LAN or WAN
5.5.0-3	The system will be connected to the agency's LAN, allowing access to all authorized users.
5.5.0-4	The system will allow access by authorized users outside the agency
5.6	**Complex coordination and controller operation**
5.6.0-1	The agency will use the following complex coordination and controller features: (Select from following needs as appropriate)
5.6.0-1.0-1	the ability to repeat a phase, such as running a left turn phase before and after its opposing through movement;
5.6.0-1.0-2	provision for the required number of rings, phases, phases per ring, and overlap phases;
5.6.0-1.0-3	the ability to operate different phase sequences based on different traffic conditions or by time-of-day;
5.6.0-1.0-4	the ability to omit a phase under some traffic conditions or based on external input to allow a shorter cycle length to operate, or to provide additional time to other phases;
5.6.0-1.0-5	special features unique to this agency, such as (give specific examples)
5.6.0-1.0-5.0-1	the ability to use flashing yellow protected/permissive and permissive only phasing
5.6.0-1.0-5.0-2	The ability to maintain coordination with external movements by preventing phases from being skipped, or by omitting phases, based on time-of-day, external input or when certain phase sequences are in operation.
5.6.0-1.0-6	The agency will permit phases or overlaps by time-of-day schedule or external input.
5.6.0-2	the ability to designate the following phases as coordinated phases; (Specify which phases may be designated as the coordinated phase)
5.6.0-3	the ability to separately monitor each lane on an approach and take different action depending on the conditions measured in each lane;
5.6.0-4	the ability to allow the coordinated phase to terminate early if the coordinated platoon is short;
5.6.0-5	the ability to introduce a non-coordinated phase later than its normal starting point within a cycle, if it can be served with minimum green within the remaining time available;
5.6.0-6	protected/permissive and permissive only phasing
5.6.0-7	support for flashing yellow protected/permissive and permissive only phasing
5.6.0-8	The agency may operate external devices using discrete signal outputs from the ASCT including occupancy on a detector, cycle length, and time-of-day. (User selects desired features.)

Con Ops Reference Number	Concept of Operations Sample Statements
5.7	5.7 Organizations Involved
6	**6 Chapter 6: Adaptive Operational Environment**
6.0-1	The system will be operated and monitored from the (specify agency) TMC.
6.0-2	The system will be operated and monitored from the (specify agency) signal shop.
6.0-3	The system will be operated and monitored from workstations located (specify who will have workstations and where they will be located).
6.0-4	An operator will be able to have full access to the system from each local controller or on-street master.
6.0-5	The central server equipment will be housed at (specify location) in an (air-conditioned or non-air-conditioned?) environment.
6.0-6	Equipment compatibility constraints
6.0-6.0-1	The central server will be a standard platform (maintained by the agency IT Department) and able to be replaced independently from the software.
6.0-6.0-2	The agency selection of controller will not be constrained by the adaptive software.
6.0-6.0-3	The agency prefers specific detector technology. (Specify your selected detector types).
6.0-6.0-4	The agency prefers to use the following controller types. (Specify acceptable controller types.)
6.0-7	The operators will already be experienced in setting up and fine tuning traditional coordinated signal systems. They will require training specific to the adaptive system, sufficient to allow them to set up, adjust and fine tune all aspects of the system.
6.0-8	The set up and fine tuning of the system will be contracted out. A review of the system's operation will be performed quarterly (specify frequency).
6.0-9	Complaints or requests for changes in operation will be handled by the in-house operators on an as-needed basis.
6.0-10	Complaints or requests for changes in operation will be handled by on-call contract staff on an as-needed basis.
6.0-11	Maintenance of all field equipment will be performed by in-house (OR contract) staff
6.0-12	Maintenance of the following field equipment will be performed by in-house (OR contract) staff. (specify what equipment will be maintained by whom)
6.0-13	Funding for maintenance of the adaptive system will come from (specify funding program or source). An increase of $xxx per year will be required to accommodate the additional equipment installed for the adaptive system.
6.0-14	Additional communications equipment and annual fees will be incurred with the adaptive system. This will amount to approximately $xxx per year, and will be covered by the (specify program or budget allocation details).
6.0-15	Replacement or repair of defective or failed equipment will be covered for xx years by the manufacturers' warranties. The labor cost of replacement during this period will be included in the purchase price.

Con Ops Reference Number	Concept of Operations Sample Statements
6.0-16	The agency expects maintenance of parts and equipment for a period of XX years will be included in the purchase price.
6.0-17	The agency expects maintenance of all adaptive system software for a period of xx years will be included in the purchase price.
6.0-18	The agency expects to operate this system using the latest software for a period of CC years.
6.0-19	The agency will seek technical support from the vendor for assistance in using the adaptive software for XX years.
6.0-20	Operations and maintenance staff will have the ability to log in to the system from remote locations via the internet, and have full functionality consistent with their access level.
6.0-21	The ASCT's operation will be able to be customized to suit the different situations that will be experienced in the different areas where it will operate.
6.0-21.0-1	The agency's experienced operators will be able to write customized routines using the ASCT's API.
6.0-21.0-2	The vendor will be able to provide customized routines that take advantage of the ASCT's API.
6.0-22	Include any additional needs for support or information from the vendor that will be needed by your agency, and that will become requirements in the contract or purchase documents.
7	**7 Chapter 7: Adaptive Support Environment**
7.1	7.1 Institutions and Stakeholders
7.1.0-1	Existing stakeholders of the traffic signal system include: (list all stakeholders, such as:) • Sponsoring agency • Neighboring agencies that operate signals • Regional agency • Fire departments • Police departments • Transit agencies • Railroad operators
7.1.0-2	The stakeholders who will be affected by or have a direct interest in the adaptive system are: (List existing and include new stakeholders)
7.1.0-3	The activities that will be undertaken by the adaptive system stakeholders include: preparation of timing parameters, implementation and fine tuning, system monitoring and adjustment, system performance monitoring and evaluation.
7.1.0-4	The organizational structures of the units responsible for installation, operation and maintenance are illustrated in the attached organization chart. The roles, responsibilities and required qualifications and experience are described below. (Describe as appropriate)
7.2	7.2 Facilities
7.2.0-1	Describe the current and/or proposed TMC.

Con Ops Reference Number	Concept of Operations Sample Statements
7.2.0-2	Will there be a satellite TMC (e.g., at Corp Yard, at a major event center, at a local EOC?)
7.2.0-3	Describe the locations elsewhere within the agency, such as on a LAN or WAN, from which access to the system will be required?
7.2.0-4	Is air-conditioning required?
7.2.0-5	Describe the location where a separate server will be located. (E.g., IT server room, TMC back room, signal maintenance area, remote hub, remote on-street cabinet)
7.2.0-6	Describe who is responsible for providing and maintaining staff facilities (e.g., personnel, public works, building services, etc.?)
7.2.0-7	Describe who is responsible for fire control facilities (e.g., part of operating group's responsibility, or the responsibility of another group, such as building services?)
7.2.0-8	Describe who is responsible for secure access to the TMC, workshop, or office with adaptive system workstations? (E.g., Is it the responsibility of the operating group or another group, such as building services?
7.3	**System Architecture Constraints**
7.3.0-1	The adaptive processor/server will be protected within the agency's firewalls. The IT Department will provide resources, equipment and system management so that operators will have appropriate access to the system locally, from within the agency's LAN and from remote locations.
7.3.0-2	The communications media available for use by the system will be: (LIST AVAILABLE MEDIA, PROVIDE A MAP OR BLOCK DIAGRAM AS APPROPRIATE. SHOW LOCATIONS OF ANY GAPS, BANDWIDTH AND LATENCY CONSTRAINTS, PROTOCOLS AND AVAILABLE ALTERNATIVES.)
7.3.0-3	The Regional ITS Architecture is illustrated in Figure XX. The adaptive system will operate within the local ITS Architecture of (NAME THE AGENCY). It will interact with the Regional ITS Architecture in the following manner. (DESCRIBE LOGICAL ARCHITECTURE AND PHYSICAL ARCHITECTURE. INCLUDE DATA FLOWS.)
7.4	**Utilities**
7.4.0-1	Are utilities the responsibility of the operating group, or are they the responsibility of another group, such as building services?
7.5	**Equipment**
7.5.0-1	Describe what test equipment is required to support the adaptive system (e.g., communications testers, fiber testers, controller testers). Is this currently available or is additional equipment required?
7.5.0-2	Will vehicles be the responsibility of the operating group or another group within the agency? What types of vehicles will be required, and how many?
7.6	**Computing hardware**
7.6.0-1	Describe the additional computing equipment required to support the operation, such as printer, copier, additional monitors, and scanner.

Con Ops Reference Number	Concept of Operations Sample Statements
7.6.0-2	Describe who is responsible for maintenance and repair of the computing equipment?
7.6.0-3	Describe who is responsible for replacement of the computing equipment when it reaches the end of its useful life?
7.7	**7.7 Software**
7.7.0-1	Who is responsible for keeping software up to date?
7.7.0-2	Who is responsible for keeping software licenses current?
7.7.0-3	What controls are proposed governing software use and availability on workstations and other support computers?
7.8	**7.8 Personnel**
7.8.0-1	Describe how many operators will be available for routine operations. Will this be provided by existing staff or will additional staff be required?
7.8.0-2	Describe what hours operators will be available.
7.8.0-3	Describe what training operators will need.
7.8.0-4	Describe what maintenance staff will be required. Will this be provided by existing staff or will additional staff be required?
7.8.0-5	What qualifications and training will the maintenance staff require?
7.9	**7.9 Operating procedures**
7.9.0-1	Describe who will be responsible for backing up databases. How often will backups be required? Will backups be stored off-site?
7.10	**7.10 Maintenance**
7.10.0-1	Describe the arrangements for maintenance. (E.g., is it done in-house or contracted out? Is it 24/7? Is equipment repair done in-house or externally?)
7.11	**7.11 Disposal**
7.11.0-1	Describe what material and/or equipment will need to be disposed of during the life of the project, and how it will be disposed.
7.11.0-2	Describe how system components will be disposed of at the end of their useful life.
8	**8 Chapter 8: Operational Scenarios**
8.1	8.1 Overview

Con Ops Reference Number	Concept of Operations Sample Statements
	The following operational scenarios describe how the system is expected to operate under various conditions. The proposed ASCT system is expected to be able to manage the following operational scenarios and issues envisioned for both the current and future project locations. Scenarios are described for the following operational conditions: (Edit to suit your situation.) • Typical heavy congested conditions • Typical heavy uncongested conditions • Moderate balanced flows • Light balanced flows • Demand affecting event • Capacity affecting event • Fault conditions (communications, detection, adaptive processor) • Signal priority and preemption • Pedestrians • Installation (For each scenario, describe the following elements: • Network • Traffic conditions • Operational objectives • Coordination and timing strategies • Summary of operations.)
8.2	**8.2 Typical Heavy (congested) Traffic**
8.2.1	*8.2.1 Example: Arterial Road with Diamond interchange*
8.2.1.1	8.2.1.1 Road network
8.2.1.1.0-1	Broadway is an arterial road that passes through a diamond interchange. While the arterial primarily provides access to the freeway from residential areas to the east and west, it also serves a major shopping area, restaurants and office land uses adjacent to the freeway. Ramp meters are used on the freeway on-ramps during periods of heavy traffic on the freeway.
8.2.1.2	8.2.1.2 Traffic conditions
8.2.1.2.0-1	During the morning peak, traffic is heavy approaching the freeway from the residential areas. Congestion occurs at the freeway interchange and two other locations (A Street and B Street). During the afternoon commuter peak, traffic is heavy departing the freeway interchange, and congestion occurs at three locations, including east-bound left turns on Broadway east of the freeway at B Street and C Street.

Con Ops Reference Number	Concept of Operations Sample Statements
8.2.1.3	**8.2.1.3 Operational objectives**
8.2.1.3.0-1	The agency has a policy of seeking smooth flow on arterial streets for routes that carry predominantly through traffic, and equitable distribution of green time at intersections that predominantly serve adjacent land uses. When congested, the agency seeks to avoid building queues on freeway off-ramps, and seeks to minimize queue spill out into through lanes. In the morning peak, the operation is designed to provide through progression approaching the freeway, and to maximize throughput at other intersections along Broadway approaching the interchange. During the afternoon peak, the operation is designed to control queue buildup on the northbound freeway off-ramp and frontage road in order to prevent queue backup onto the freeway. The operational objectives under these conditions are to: - Accommodate the traffic at all intersections with a minimum of phase failures - Control inflows to the diamond interchange to prevent queue spillback into upstream intersections; and - Provide smooth flow along the arterial road.
8.2.1.4	**8.2.1.4 Coordination and signal timing strategies**
8.2.1.4.0-1	The diamond interchange runs a TTI-four-phase operation from a single signal controller. The coordination approach for the morning peak is progression, maximizing bandwidth in the direction approaching the freeway interchange. This requires a 90-second cycle which provides good resonant progression in both directions, with 40% bandwidth efficiency in the peak direction, when side-street volumes are low. This cycle length also minimizes delay with no effect on throughput at the diamond, given that the lost time is offset by the internal double clearance, which means that increasing the cycle length does not increase throughput. The coordination approach in the afternoon peak is to maximize progression bandwidth leaving the interchange, except at X which is routinely congested and the agency seeks to allow queues to build on the side-street approach to maximize throughput on the arterial. Queue formation on the eastbound arterial approach to the freeway is allowed to maximize the green time and throughput for the northbound ramp approach. The signal timing strategies used by the system to accommodate this situation are: - At the diamond interchange, select phase times that ensure queues do not exceed storage lengths. - At the critical intersection(s), select phase sequence that eliminates queue overflow in left turn bays - At each intersection, select phase times that eliminate phase failures - At the other arterial road intersections, provide sufficient time to serve all turning and side street traffic without phase failures - At the other arterial road intersections, provide green on the coordinated route phases in a manner that minimizes the stops for through traffic along the arterial.
8.2.1.5	**8.2.1.5 Summary of Operation**

Con Ops Reference Number	Concept of Operations Sample Statements
8.2.1.5.0-1	The actuated system will measure the traffic flow and determine when each of these operational objectives should be in force, and therefore which of the coordination and timing strategies to give priority to in making its adaptive decisions. The adaptive system will use the 90-second cycle in the morning peak to preserve resonant progression. The adaptive system will not alter the operation of the diamond interchange phase sequence. The adaptive system will seek to balance green time utilization when side-street demand is important, such as during the noon peak. The adaptive system will seek to minimize residual queuing at congested locations, preferring to build queue at x and y if demand cannot be accommodated. The adaptive system will prevent residual queue buildup on the freeway ramps. The adaptive system reports bandwidth, arrivals on red as a measure of bandwidth utilization, and phase utilization measurements that were used to adaptively adjust green times.
8.2.2	*8.2.2 Example: Arterial with one critical intersection*
8.2.2.1	8.2.2.1 Road network
8.2.2.1.0-1	The section of Broadway Road to be coordinated using ASCT has six signalized intersections. It is a six lane arterial road with a two way left turn lane, and exclusive left turn lanes at each intersection. Most of the intersections provide access to local businesses and residential areas. However, one intersection (name of cross street) is an arterial road that accommodates regional traffic rather than providing local access. There are no nearby signals on this cross street that require coordination with this critical intersection. This is an eight-phase intersection with protected left turns on all approaches. The other intersections have permissive left turns on the side streets. Broadway is classified by the MPO as an arterial road of regional significance.
8.2.2.2	8.2.2.2 Traffic conditions
8.2.2.2.0-1	There is one critical intersection (Cross Street) that has heavier traffic than the other intersections at all times of the day. At its heaviest (typically during the AM and PM peaks) most movements are congested with occasional phase failures. Traffic is heaviest in one direction when these conditions are experienced, typically northbound during the AM peak and southbound during the PM peak. The traffic on Broadway is 50% heavier than the traffic on (Cross Street) during this condition.
8.2.2.3	8.2.2.3 Operational objectives
8.2.2.3.0-1	The operational objectives for this arterial under these conditions are to: • Maximize the throughput along Broadway; • Accommodate the traffic at the critical intersection with a minimum of phase failures; and • Provide smooth flow along the arterial through other intersections.
8.2.2.4	8.2.2.4 Coordination and signal timing strategies

Con Ops Reference Number	Concept of Operations Sample Statements
8.2.2.4.0-1	The signal timing strategies used by the system to accommodate this situation are: • At the critical intersection, select phase times that eliminate phase failures • At the critical intersection, select phase sequence that eliminates queue overflow in left turn bays • At the critical intersection, select phase times that eliminate queue overflow in left turn bays • At the critical intersection, distribute green time to maximize the throughput on Broadway • At the non-critical intersections, provide sufficient time to serve all turning and side street traffic without phase failures
8.2.2.5	8.2.2.5 Summary of operation
8.2.2.5.0-1	Under these conditions, the ASCT system will select a phase arrangement and calculate phase times that accommodate traffic at the critical intersection. It will then set the timing at the other intersections to provide a green band in the direction of heaviest traffic along the arterial, to minimize the number of stops in that direction. The green time for the non-arterial phases at those intersections will be set to accommodate the traffic using those phases, while allocating the remaining time to the arterial road. The system will determine the sequence of phases on the arterial (lead-lead, lead-lag or lag-lag) that minimizes the stops in the non-coordinated direction under these conditions.
8.2.3	8.2.3 *Example: Arterial with several critical intersections*
8.2.3.1	8.2.3.1 Road network
8.2.3.1.0-1	The section of Broadway Road to be coordinated using ASCT has ten signalized intersections. It is a six lane arterial road two way left turn lane, and exclusive left turn lanes at each intersection. Two intersections (name of cross streets) are arterial roads that accommodate regional traffic rather than providing local access. There are no nearby signals on the cross streets that require coordination with this critical intersection. One intersection provides access to a major shopping district. These are all eight-phase intersections with protected left turns on all approaches. The remaining intersections provide access to local businesses and residential areas. Those intersections have protected left turns on Broadway and permissive left turns on the side streets. Broadway is classified by the MPO as an arterial road of regional significance.
8.2.3.2	8.2.3.2 Traffic conditions
8.2.3.2.0-1	At times when traffic conditions are very heavy, one of the three key intersections is the critical intersection. This varies depending on the level of demand on the two crossing arterials or activity in the shopping district. When traffic is very heavy, it is typically heaviest on Broadway in one direction (such as northbound during the AM peak and southbound during the PM peak). In these conditions, Broadway carries higher volumes than the crossing arterials.
8.2.3.3	8.2.3.3 Operational objectives
8.2.3.3.0-1	The operational objectives for this arterial under these conditions are to: • Maximize the throughput along Broadway • Accommodate the traffic at the critical intersection with a minimum of phase; and • Provide smooth flow along the arterial through other intersections.

Con Ops Reference Number	Concept of Operations Sample Statements
8.2.3.4	8.2.3.4 Coordination and signal timing strategies
8.2.3.4.0-1	The signal timing strategies used by the system to accommodate this situation are: • Determine the critical intersection • At the critical intersection, select phase times that eliminate phase failures • At the critical intersection, select phase sequence that eliminates queue overflow in left turn bays • At the critical intersection, distribute green time to maximize the throughput on Broadway. • At the non-critical intersections, provide sufficient time to serve all turning and side street traffic without phase failures • At the non-critical intersections, provide green on the arterial road phases in a manner that minimizes the stops for through traffic along the arterial.
8.2.3.5	8.2.3.5 Summary of operation
8.2.3.5.0-1	Under these conditions, the ASCT system will determine the critical intersection and select a phase arrangement and calculate phase times that accommodate traffic at that intersection. It will then set the timing at the other intersections to provide a green band in the direction of heaviest traffic along the arterial, to minimize the number of stops in that direction. The green time for the non-arterial phases at those intersections will be set to accommodate the traffic using those phases, while allocating the remaining time to the arterial road. The system will determine the sequence of phases on the arterial (lead-lead, lead-lag or lag-lag) that minimizes the stops in the non-coordinated direction under these conditions.
8.2.4	8.2.4 *Example: Crossing arterials*
8.2.4.1	8.2.4.1 Road network
8.2.4.1.0-1	Broadway is an arterial road with seven signalized intersections. Cross Street is also an arterial road with five signalized intersections, and it crosses Broadway, as illustrated in the figure.
8.2.4.2	8.2.4.2 Traffic conditions
8.2.4.2.0-1	During heavy traffic conditions (such as AM and PM peak) the Broadway/Cross Street intersections is the critical intersection, and queues develop on all approaches. Typically the northbound direction on Broadway is significantly heavier than the southbound. Likewise, the eastbound traffic on Cross Street is significantly heavier that the westbound.
8.2.4.3	8.2.4.3 Operational objectives
8.2.4.3.0-1	The operational objectives for these arterials under these conditions are to: • Maximize the throughput along Broadway • Maximize the throughput along Cross Street • Accommodate the traffic at the critical intersection with a minimum of phase failures; and • Provide smooth flow along the arterial through other intersections.
8.2.4.4	8.2.4.4 Coordination and signal timing strategies

Con Ops Reference Number	Concept of Operations Sample Statements
8.2.4.4.0-1	The signal timing strategies used by the system to accommodate this situation are: • At the critical intersection, select phase times that eliminate phase failures • At the critical intersection, select phase sequence that eliminates queue overflow in left turn bays • At the non-critical intersections on both arterials, provide sufficient time to serve all turning and side street traffic without phase failures • At the non-critical intersections, provide green on the arterial road phases in a manner that minimizes the stops for through traffic along the arterial.
8.2.5	*Example: Grid network*
8.2.5.1	Road network
8.2.5.1.0-1	The intersections to be coordinated are on a grid network with relatively fixed intersection spacing. The roads typically have four lanes plus separate left turn bays at intersections. Based on the intersection spacing and typical mid-block vehicle speeds, there is a resonant cycle length of approximately 60 seconds that would provide coordination on most streets. This cycle length would also accommodate pedestrian movements at all intersections.
8.2.5.2	Traffic conditions
8.2.5.2.0-1	During heavy traffic conditions (such as peak shopping periods and PM peak hours), the demand at many of the intersections cannot be accommodated at the resonant cycle length. In addition, at key locations the block length is such that not all of the demand on some approaches can be stored in the approach block.
8.2.5.3	Operational objectives
8.2.5.3.0-1	The operational objectives for the streets in this network under these conditions are to: • Accommodate the traffic at all intersections with a minimum of phase failures; • Control inflows to blocks to prevent queue spillback into upstream intersections; and • Provide smooth flow along as many routes as possible through the network.
8.2.5.4	Coordination and signal timing strategies
8.2.5.4.0-1	The signal timing strategies used by the system to accommodate this situation are: • Determine which routes need to be coordinated and which blocks can best accommodate no coordination, • At each intersection, select phase times that minimize phase failures; • Determine the critical intersection(s) within the network • At the critical intersection(s), select phase sequence that minimizes queue overflow in left turn bays • At the critical intersection(s), distribute green time to maximize the throughput on the coordinated routes. • At the non-critical intersections, provide sufficient time to serve all turning and side street traffic without phase failures • At the non-critical intersections, provide green on the coordinated route phases in a manner that minimizes the stops for through traffic along the coordinated route. • At intersections with limited approach block length, set the timing of upstream intersections so that queues do not exceed the block length.

Con Ops Reference Number	Concept of Operations Sample Statements
8.2.5.5	8.2.5.5 Summary of Operation
8.2.5.5.0-1	The network will operate at a 60 second cycle length, because that has been determined to be a resonant cycle length at which two-way progression can be provided on the coordinated routes. The ASCT will select the most appropriate phase sequence at intersections where phase sequence is permitted to vary, and select phase times that accommodate all pedestrian activity and distribute green times to minimize phase failures, and implement offsets that provide progression along the coordinated routes. Because XX block is short, the offsets will be set so that when a coordinated platoon passes through A Street, it will always clear B Street, so the block is cleared and available to store turning traffic from A Street.
8.3	8.3 Typical Heavy (Uncongested) Traffic
8.3.1	8.3.1 Example: Isolated Intersection
8.3.1.1	8.3.1.1 Road network
8.3.1.1.0-1	A Road and B Road are two important limited access arterials that intersect and there are no other intersections sufficiently close that traffic flow would benefit from providing coordination. At the intersection, each road has three through lanes on each approach, dual left turn lanes and exclusive right turn lanes. Although pedestrian crosswalks are provided, there are rarely pedestrians at this isolated location.
8.3.1.2	8.3.1.2 Traffic conditions
8.3.1.2.0-1	Traffic is heavily directional during the commuter peaks. A Road is predominantly northbound during the AM and southbound during the PM, while B Road is predominantly eastbound during the AM and westbound during the PM. There is significant turning traffic in the peak directions. The left turn bays in the peak direction often overflow (east to north during the AM peak and west to south during the PM peak). There are occasional phase failures resulting in carryover queues at the end of phases. However, because of the high volumes and relatively long queues that form under vehicle-actuated (free) operation, there is a significant portion of each phase green during which the throughput is well below saturation flow, but not sufficiently low that phase gap-out occurs. The intersection delay (and therefore the LOS) would be improved by using a lower cycle length than can be achieved using normal vehicle-actuated operation.
8.3.1.3	8.3.1.3 Operational objectives
8.3.1.3.0-1	The operational objective for this case is to reduce delay by improving the efficiency of each phase.
8.3.1.4	8.3.1.4 Coordination and signal timing strategies
8.3.1.4.0-1	The signal timing strategies used by the system to accommodate this situation are: • Select a cycle length that minimizes overall delay at the intersection • Select a phase sequence that maximizes the efficiency of the movements in the peak direction • Distribute phase times to minimize phase failures on all approaches • Modify cycle length and phase times if necessary to accommodate occasional pedestrians
8.3.1.5	8.3.1.5 Summary of Operation

Con Ops Reference Number	Concept of Operations Sample Statements
8.3.1.5.0-1	The adaptive system will measure the traffic flow and determine the appropriate cycle length and phase times to accommodate the current demand. When traffic volumes are sufficiently high, lead-lag operation will be selected for one or both approaches and unused time generally added to the phases serving the peak directions.
8.4	**Moderate balanced flows**
8.4.1	*Arterial road with irregular spacing*
8.4.1.1	Road network
8.4.1.1.0-1	The section of Broadway Road to be coordinated using ASCT has six signalized intersections. It is a six lane arterial road with a two way left turn lane, and exclusive left turn lanes at each intersection. Most of the intersections provide access to local businesses and residential areas. However, one intersection (Cross Street) is an arterial road that accommodates regional traffic rather than providing local access. There are no nearby signals on this cross street that require coordination with this critical intersection. This is an eight-phase intersection with protected left turns on all approaches. The other intersections have permissive left turns on the side streets. There is no regular spacing between the intersections and therefore no "resonant" cycle length. Broadway is classified by the MPO as an arterial road of regional significance.
8.4.1.2	Traffic conditions
8.4.1.2.0-1	During business hours traffic uncongested and the flows along Broadway are similar in both directions. At lunch time there is an increase in traffic turning into and out of the several side streets that service local shops and restaurants. There is little pedestrian activity except at Cross Street where there are bus stops and local shops. There is enough side street and turning movement traffic that most signal phases are called every cycle. The left turn volumes are sufficiently high that they need protected turn phases to provide sufficiently capacity and prevent phase failures.
8.4.1.3	Operational objectives
8.4.1.3.0-1	The operational objectives for this condition are to: • Provide smooth flow along Broadway; and • Provide signal timing that prevents phase failures at all intersections.
8.4.1.4	Coordination and signal timing strategies
8.4.1.4.0-1	The coordination approach for these conditions is provide progression, maximizing bandwidth while providing two-way coordination. This can be done at a resonant cycle length of 80 seconds. The strategies applied while maintaining this cycle length are: • At each intersection, provide sufficient time to serve all turning and side street traffic without phase failures; • At each intersection, select phase times (or offsets) that provide smooth flow along the arterial in both directions. • At each intersection, select phase sequence that provides smooth flow along the arterial • At the specified intersection, select phase times that will accommodate frequent use of all pedestrian phases. • At other intersections, select phase times that will accommodate occasional use of pedestrian phases.

Con Ops Reference Number	Concept of Operations Sample Statements
8.4.1.5	8.4.1.5 Summary of Operation
8.4.1.5.0-1	The critical intersection will determine the minimum cycle length that can be used for the entire group. This cycle length will accommodate all phases and all pedestrian movements. Provided it is not higher than the 90 second resonant cycle length, the system will set the cycle length to be 90 seconds. It will detect the balanced flows and select offsets that provide a reasonable compromise between the two directions of travel. At the non-critical intersections, the non-coordinated phases will be set to accommodate pedestrians and vehicles, and all spare time will be allocated to the coordinated phases to maximize the bandwidth for progression bands along the road. During periods (such as lunch time) when there is more turning traffic associated with local retail activity) extra time will be provided to those phases within the overall cycle length, at the expense of the coordinated phases on Broadway
8.5	8.5 Light Balanced Flows
8.5.1	8.5.1 Arterial Road
8.5.1.1	8.5.1.1 Road network
8.5.1.1.0-1	The section of Broadway Road to be coordinated using ASCT has six signalized intersections. It is a six lane arterial road with a two way left turn lane, and exclusive left turn lanes at each intersection. Most of the intersections provide access to local businesses and residential areas. However, one intersection (Cross Street) is an arterial road that accommodates regional traffic rather than providing local access. There are no nearby signals on this cross street that require coordination with this critical intersection. This is an eight-phase intersection with protected left turns on all approaches. The other intersections have permissive left turns on the side streets. Broadway is classified by the MPO as an arterial road of regional significance.
8.5.1.2	8.5.1.2 Traffic conditions
8.5.1.2.0-1	During some periods of the weekdays and weekends traffic is light but predominantly passing along the arterial. There is little pedestrian activity and little side street and turning movement traffic. The left turn volumes are sufficiently light that they do not need protected turn phases to provide sufficiently capacity, and can normally be accommodated by permissive phases. There is a resonant cycle length of 45 seconds that will provide two-way coordination when the protected left turn phases are omitted.
8.5.1.3	8.5.1.3 Operational objectives
8.5.1.3.0-1	The operational objectives for this condition are to
8.5.1.3.0-2	• Provide smooth flow along Broadway; and
8.5.1.3.0-3	• Provide signal timing that prevents phase failures at all intersections.
8.5.1.4	8.5.1.4 Coordination and signal timing strategies

Con Ops Reference Number	Concept of Operations Sample Statements
8.5.1.4.0-1	The coordination approach for this condition is to provide progression along the arterial, maximizing bandwidth while providing two-way coordination. The timing strategies applied to do this are: • At each intersection, provide sufficient time to serve all turning and side street traffic without phase failures; • At each intersection, select phase times (or offsets) that provide smooth flow along the arterial in both directions; • At each intersection, omit protected turning phases to minimize the impact of occasional turning vehicles on other traffic; • At each intersection, select phase times that will accommodate occasional use of pedestrian phases.
8.5.1.5	8.5.1.5 Summary of Operation
8.5.1.5.0-1	During light traffic conditions, protected left turn phases will be omitted and a cycle length of 45 seconds implemented. If traffic volumes decrease (such as during late nights), the 45 second cycle length and two-way coordination will be maintained unless the volumes fall below a minimum threshold, in which case the signals will be set to operate in free, vehicle-actuated mode. If traffic increases to the extent that it can no longer be accommodated within the 45 second cycle length, or left turning volumes can no longer be accommodated without the protected left turns, then a longer cycle length will be implemented and a new coordination strategy selected to match the current traffic conditions.
8.6	8.6 Demand affecting event
8.6.1	8.6.1 *High travel day (e.g., Mothers' Day, Superbowl)*
8.6.1.0-1	During periods of major activity within or close to the ASCT's area of operation, the traffic characteristics are often similar to the peak periods, either oversaturated or unsaturated. The system will behave in a similar fashion to those periods, and the detection system will determine whether unsaturated or oversaturated conditions prevail. If there is heavily directional traffic before or after the activity, the system will determine the predominant direction and coordinate accordingly, with an appropriate cycle length and offset. If the event traffic is not as heavy as peak hours, but the traffic on the corridor is still highly directional, then the system will recognize this and provide coordination predominantly in the heaviest direction, even though the cycle length may be similar to business hours (with balanced flows) cycle lengths. The entire corridor may be set by the operator to operate as one or more coordinated groups under this condition, or the system may have the freedom to operate it as one or more groups subject to user-specified criteria, such as similar required cycle lengths in different parts of the corridor, or the volume of traffic at key locations exceeds a threshold.
8.7	8.7 Capacity affecting event
8.7.1	8.7.1 *Weather event*
8.7.2	8.7.2 *Incident within the system (construction, maintenance, fire)*

Con Ops Reference Number	Concept of Operations Sample Statements
8.7.2.0-1	When an incident occurs on the coordinated route and temporarily reduces the capacity of the route (such as emergency vehicles stopped, unscheduled construction/maintenance, or traffic crash), there will typically be congestion upstream of the blockage, and lighter than normal traffic downstream. In such a situation, it is appropriate for the downstream signals to operate with different characteristics from the upstream signals. If the downstream signals experience lighter traffic as a result of the blockage, those signals should be coordinated as a group, with cycle length, splits and/or offsets that react to the measured traffic. If the blockage is in the peak direction, then it may be appropriate to coordinate in the opposite direction if that traffic is similar to or greater than the normal peak direction. If the blockage is in the non-peak direction, there may be no need to depart from the normal operation. While intersections upstream from the blockage may register increased congestion, the appropriate response would not be to increase the capacity in the congested direction. On the contrary, the approach should be to match the capacity for phases in the direction towards the bottleneck to the actual capacity of the bottleneck, and prevent this movement from adversely affecting cross street traffic and the flow in the non-affected direction. The system will recognize the presence of an abnormal obstruction and modify the signal operation to react to the changed traffic conditions in an efficient manner.
8.8	**8.8 Fault Conditions**
8.8.1	*8.8.1 Communications Fault Condition*
8.8.1-1	If a communication failure prevents the adaptive system from continuing to control one or more intersections within a defined group, all signals within the group will revert to an appropriate, user-specified fallback mode of operation, either time-of-day operation or free operation. The fallback mode will be specified by the user based on location and time of day. All communication failure alarms will be automatically transmitted to maintenance and operations staff for appropriate attention.
8.8.2	*8.8.2 Detection Fault Condition*
8.8.2.0-1	The system will recognize a detector failure and take appropriate action to accommodate the missing data. For a local detector failure, the local controller will place a soft recall or maximum recall (to be user-specified) on the appropriate phase, and issue an alarm. For a detector that influences the adaptive operation (e.g., a system detector), the system will use data from an alternate (user-specified) detector, such as in an adjacent lane or at an appropriate upstream or downstream location. If the number of detector failures within a specified group exceeds a user-specified threshold, the system will cease adaptive operation and go to a fallback operation specified by the user (such as time-of-day operation or free operation). The fallback operation will be specified by the user based on location and time of day. All detector failure alarms will be automatically transmitted to maintenance and operations staff for appropriate attention.
8.9	**8.9 Priority and Preemption**
8.9.1	*8.9.1 Railroad Preemption*
8.9.1.1	8.9.1.1 Example Preemption Scenario - Example 1

Con Ops Reference Number	Concept of Operations Sample Statements
8.9.1.1.0-1	XX arterial runs north-south and there are gated railroad grade crossings on several of the east-west routes that cross XX Arterial, namely XX Blvd and XX St. The rail line is approximately XXft. to the east of XX Arterial. The railroad gated crossing preempts the signals at XX intersection and also XX (TWO INTERSECTIONS ON THE ARTERIAL). Upon preemption, the signals on XX arterial introduce a clearance phase, to ensure any vehicles queued on or close to the railroad tracks can clear before the gates descend. Upon completion of the clearance interval, the signal continues limited operation. The phases that would normally send traffic in the direction of the grade crossing are inhibited until the gates are raised. Once the clearance sequence is completed, the signal returns to normal operation. There is also a queue detector on the eastbound departure side of XX Arterial, which detects the presence of a queue approaching the grade crossing when the gates are lowered. If such a queue is detected, the phases that normally send traffic in the direction of the grade crossing are inhibited as long as the gates remain lowered. When an intersection responds to railroad preemption, all signals within the coordinated group are released to local control, and operate according to a time-of-day schedule. Once the preemption is released, all the signals in the coordinated group return to adaptive control.
8.9.2	*8.9.2 Light Rail Priority*
8.9.2.1	8.9.2.1 Example Light Rail Priority Scenario.
8.9.2.1.0-1	LRT priority will be provided at each intersection on an LRT route. The input requesting priority will come EITHER from the centralized priority system The system will have the capability to extend the existing green if that will serve the LRV, introduce an early green by shortening or skipping other phases, or run a phase called exclusively by the LRV.

The decision to provide priority will be determined within the local controller, based on user-definable and settable rules. These rules will include such items as: length of time or number of cycles since last priority was provided, and priority level if there are competing requests.

The LRT system has its own logic to determine whether a priority request for an approaching LRV will be transmitted to the signal controller, based on such parameters as schedule adherence, route number, in-service or out-of-service and passenger loading. This logic will not reside within the adaptive system. |
| 8.9.3 | *8.9.3 Bus Signal Priority* |
| 8.9.3.1 | 8.9.3.1 Example Bus Priority Scenario |
| 8.9.3.1.0-1 | Bus priority will be provided at each intersection on a bus route. The input requesting priority will come from the centralized priority system.

The system will have the capability to extend the existing green if that will serve the bus, introduce an early green by shortening or skipping other phases, or run a phase called exclusively by the bus.

The decision to provide priority will be determined within the local controller, based on user-definable and settable rules. These rules will include such items as: length of time or number of cycles since last priority was provided, and priority level if there are competing requests.

The bus system has its own logic to determine whether a priority request for an approaching bus will be transmitted to the signal controller, based on such parameters as schedule adherence, route number, in-service or out-of-service and passenger loading. This logic will not reside within the adaptive system. |

Con Ops Reference Number	Concept of Operations Sample Statements
8.9.4	*Emergency Vehicle Preemption*
8.9.4.0-1	When an intersection responds to an EV preemption, other signals within the coordinated group continue to operate adaptively. The preempted signal returns to adaptive control once the preemption is released
8.10	**Scheduled Events**
8.10.0-1	The system will recognize the increasing traffic as patrons arrive for the event and adopt an appropriate mode of operation. During the event, when there is little associated traffic, the system will recognize the traffic conditions and operate normally, then recognize the changing traffic pattern as patrons begin to leave the event and adopt the appropriate mode of operation until the traffic clears. The system will then return to normal operation.
8.11	**Pedestrians**
8.11.0-1	Pedestrian crossing times must be accommodated. At locations with wide pedestrian crosswalks and a history of conflicts between turning vehicles and pedestrians, the pedestrian walk is displayed some seconds before the compatible vehicle green. At crosswalks with high pedestrian volumes, a pedestrian recall is used during the periods when the pedestrian volumes are high. Pedestrian recall is used for pedestrian phases that are adjacent to the coordinated movements. During periods when pedestrian volumes are high and queuing of the conflicting right turn movement becomes unacceptable, the vehicles are directed elsewhere by prohibiting the movement (such as by operating a No Right Turn sign). When side street traffic is light and no pedestrian is present, a vehicle may arrive on the side street shortly after the point at which its phase would normally be initiated. Typically it would then wait an entire cycle before being served. However, it is often possible to serve one or two side street vehicles within the remaining green time. So the system will be able to start a phase later than normal when there is no pedestrian call for that phase, provided it can be completed before the time the phase would normally end.
8.12	**Installation**
8.12.0-1	During installation and fine tuning, the operator will calibrate all the user-defined values in the system. In order to understand the response of the system to changes in traffic conditions, it is necessary to examine the results of intermediate calculations, in addition to the overall outputs and changes of state commanded by the system. For example, if a cycle length is calculated based on a calculated parameter, such as level of saturation of detectors in critical lanes on critical movements, then the state of that calculated parameter must be available for inspection for each detector. This will allow the operator to properly calibrate each detector, and then separately calibrate the parameters in the cycle length calculation or look-up table. This would also allow an operator to identify a faulty detector that is

Appendix C
Concept of Operations
Examples Scenarios

OPERATIONAL SCENARIOS

These operational scenarios have been extracted from a Concept of Operations prepared for a real project. The street names and place names have been changed.

INITIAL CORRIDOR

This section describes operational scenarios envisioned for the pilot deployment in the Broadway Avenue corridor. The wider deployment of the adaptive system will involve some additional operational characteristics that will not be applicable to the pilot project. Several extended scenarios are described in a later section. Although the term "cycle length" is used in the following descriptions, this should not necessarily be interpreted to mean a constant, repeating cycle length. It may represent a variable length of time between the start of a phase sequence, using a fixed sequence, or also the length of time between the start of a particular phase (say a nominated coordinated phase) with a variable sequence of phases occurring between consecutive occurrences of the reference point.

PEAK PERIODS – UNSATURATED CONDITIONS

During typical peak periods (and other periods when traffic volumes are high), the system will select a cycle length that accommodates all movements at all intersections. The primary determinant of the cycle length will be to ensure that there are no phase failures on critical movements that would adversely affect the operation of other intersections or the progress in the peak direction. It is expected that the cycle length will vary according to the traffic conditions, increasing as the peak period begins and decreasing as it dies down.

The system will compare the volumes traveling in each direction, and provide coordination in the dominant direction. Should the volumes be balanced, the coordination will be implemented in a manner that provides balanced progression as far as possible in the two directions.

Where leading and lagging left turn phases are used, the system will determine the optimal phase sequence in order to provide the best coordination. This would be linked to the direction of offset, such as providing a lagging left turn in the heavy, coordinated direction. If the green time required for a left turn phase is longer the time required to service a queue fully occupying the left turn bay, and the queue would overflow and block the adjacent lane, the operator will be able to specify the phase to operate twice per cycle in order to avoid queue overflow.

The entire corridor may be set by the operator to operate as one coordinated group, or the system may have the freedom to operate it as one group subject to user-specified criteria, such as similar required cycle lengths in different parts of the corridor are similar or the volume of traffic in the peak direction exceeds a threshold.

PEAK PERIODS – OVERSATURATED CONDITIONS

During peak periods when one or more intersections are oversaturated, the primary objective of the system will be to maximize the throughput along the corridor in the peak direction. The cycle length chosen by the system will be the maximum permitted by the operator, or determined by a user-specified maximum duration between successively servicing a phase with demand present. The system will determine the direction with peak flow and provide the maximum bandwidth possible within the selected cycle length. This will be subject to user-specified constraints, such as allowable phase sequences, and minimum and maximum phase times.

As described in the unsaturated peak description, phase sequence of lead-lag phases, and the operation of left turn phases twice per cycle, will be determined by the system. The entire corridor may be set by the operator to operate as one coordinated group, or the system may have the freedom to operate it as one group subject to user-specified criteria, such as similar required cycle lengths in different parts of the corridor are similar or the volume of traffic in the peak direction exceeds a threshold.

BUSINESS HOURS

During business hours, there will be two separate but complementary objectives: select a cycle length that ensures all movements at all intersections are accommodated equitably, while providing reasonable coordination in one or both directions. Because of the block spacing on Broadway Avenue and the associated travel times between the key intersections, there are several cycle lengths that provide good coordination in both directions. Provided a cycle length that provides good coordination exceeds the minimum cycle length needed to serve all movements at all intersections, the system will pick that cycle length. During this period, there are sufficient pedestrians present at the key intersections that the cycle length will need to accommodate pedestrians on most phases in almost every cycle. If the demand peaks (as sometimes happens during lunchtime), the system will increase the cycle length and select the appropriate progression in the same manner as during unsaturated peak conditions.

The entire corridor may be set by the operator to operate as one or more coordinated groups under this condition, or the system may have the freedom to operate it as one or more groups subject to user-specified criteria, such as similar required cycle lengths in different parts of the corridor are similar or the volume of traffic at key locations exceeds a threshold.

OFF-PEAK PERIODS

During early mornings, evenings and parts of the weekends when traffic is lighter than during the business hours, the coordination objectives will be similar to the business hours, although a lower cycle length may be applicable. If there is a cycle length that would provide good two-way progression and accommodate all movements at all intersections equitably, but cannot accommodate all pedestrian movements on all phases and stay in coordination, the system will allow the lower cycle length through the following actions. If protected/permitted left turn phasing is in operation, the protected phase can be omitted under user-specified conditions, such as very light volume or short queue lengths (determined by detector logic). The maximum green time may be set lower than the sum of pedestrian walk and clearance times, and still allow the pedestrian phase to operate by extending the green time when necessary without throwing the system out of coordination.

During normal weekend traffic conditions, the system may operate in the same manner as the business hours or as the off-peak periods.

The entire corridor may be set by the operator to operate as one or more coordinated groups under this condition, or the system may have the freedom to operate it as one or more groups subject to user-specified criteria, such as similar required cycle lengths in different parts of the corridor are similar or the volume of traffic at key locations exceeds a threshold.

LRT PREEMPTION

LRT operation is an extremely important part of the corridor operation. There are gated grade crossings of the light rail line on several of the east-west routes that cross Broadway Avenue, namely Northern, E. Central Blvd and E. Southern St. The light rail line is approximately 600ft. to the east of Broadway, and there is a signalized intersection with S. Telegraph Avenue approximately 400ft. east of Broadway. The light rail preempts the signals at S. Telegraph Avenue and also preempts the S. Arroyo Parkway signals (operated by State DOT) to the east of the light rail line. The light rail currently provides advance warning of an approaching train 72 seconds before its expected arrival at the crossing. This provides time for the controller to cycle through appropriate clearance phases (such as eastbound through and left turn phases at Arroyo) before entering the preemption operation (either flashing red or limited service, depending on the location).

There are times when the westbound queuing is sufficiently heavy that preemption of the Broadway signals may be necessary in order to clear the tracks safely. A preemption call received at S. Telegraph will be transmitted to the controller at S. Broadway, and logic applied so that a preemption phase will be able to be initiated if there is insufficient queuing space available in the block between S. Telegraph and S. Broadway to store vehicles that clear the block between S. Telegraph and the light rail line. This logic will be determined by a queue detector west of S. Telegraph in the westbound lanes. This is expected to result in preemption generally not being required outside the peak periods, and only being required within the peak periods when queuing is present on the westbound approach to Broadway.

The existing arrangement of inhibiting the eastbound through phase at S. Broadway during a preemption if the eastbound block is full will be continued. A queue detector east of S. Broadway in the eastbound direction detects the presence of a queue. If the gates are down, the eastbound detectors are suppressed, and the eastbound through phase does not run.

Once the gates are raised, the signals enter a post-preemption sequence. Typically this involves serving the east-west phases first to clear or reduce the queus on those movements that built up during the preemption.

BUS PRIORITY

Bus priority will be provided at each intersection. The input requesting priority will come either from the centralized METRO priority system or directly from the approaching bus, depending on location and bus route. The system will have the capability to extend the existing green if that will serve the bust, introduce an early green by shortening or skipping other phases, or run a phase called exclusively by the bus. The decision to provide priority will be determined within the local controller, based on user-definable and settable rules. These rules will include such items as: length of time or number of cycles since last priority was provided, and priority level if there are competing requests. It is anticipated that the various bus priority systems that will be in place will have their own logic to determine whether a priority request for an approaching bus will be transmitted to the signal controller, based on such parameters as schedule adherence, route number, in-service or out-of-service and passenger loading. This logic will not reside within the adaptive system.

EMERGENCY VEHICLE PREEMPTION

When a fire call is received, a request for a preemption route will be placed from the fire house. This will be achieved by either activating a switch inside the fire house or by placing a preemption call to the first intersection from the fire truck. If there are multiple routes from which the fireman must choose one, there would generally by multiple switches inside the fire house. Preemptions will be placed at the appropriate intersections along the route. If travel time to an intersection is longer than required for any clearance period, the introduction of the preemption will be delayed until it is required, and any potentially conflicting vehicle or pedestrian phases that would delay the preemption will be suppressed.

At other intersections beyond the pre-set routes, emergency vehicle preemption will occur in the normal fashion when requested by an approaching emergency vehicle.

MAJOR EVENTS

During major events, the traffic characteristics are often similar to the peak periods, either oversaturated or unsaturated. The system will behave in a similar fashion to those periods, and the detection system will determine whether unsaturated or oversaturated conditions prevail. If there is heavily directional traffic before or after an event, the system will determine the predominant direction and coordinate accordingly, with an appropriate cycle length and offset. If the event traffic is not as heavy as peak hours, but the traffic on the corridor is still highly directional, then the system will recognize this provide coordination predominantly in the heaviest direction, even though the cycle length may be similar to business hours (with balanced flows) cycle lengths.

The entire corridor may be set by the operator to operate as one or more coordinated groups under this condition, or the system may have the freedom to operate it as one or more groups subject to user-specified criteria, such as similar required cycle lengths in different parts of the corridor are similar or the volume of traffic at key locations exceeds a threshold.

MAJOR INCIDENTS

When a major incident occurs on one of the freeways, or at a location within Central City, the traffic on Broadway Avenue will change in a manner that is difficult to predict, and the response required of the system will vary depending on the time of day, day of week and the current traffic conditions at the time the incident occurs. The system will detect any increase in traffic volume and take the following action. If the increased volume needs a higher cycle length in order to continue to accommodate all movements at all intersections, it will increase the cycle length, but only up to the maximum permitted by the operator. If the diverted traffic results in a change in the balance of the direction of the traffic on the corridor, the progression will be changed to match the traffic. Typically the result of these actions will be to increase the cycle length and provide a wide progression bandwidth in the direction of the diverted traffic. However, if the incident occurs at times of lower overall traffic volumes, and it does not result in oversaturated conditions on the corridor, the result may be that the system mimics a typical peak pattern or business hours pattern.

This type of incident will typically not result in a uniform increase in traffic in one direction for the entire length of the corridor. If traffic diverts from one of the freeways to the north or south of this corridor, it often will turn onto one or more of the important east-west corridors. Therefore, it is expected that the response of the system will be different in the northern and southern parts of the corridor, depending on the location, nature and time of day of the incident. The architecture of the system will allow the northern, central and southern portions of the system to respond independently but in a consistent manner during incidents.

DETECTOR FAILURE

Detector reliability is a very important part of successful adaptive operation. The system will recognize a detector failure and take appropriate action to accommodate the missing data. For a local detector failure, the local controller will place a soft recall or maximum recall (to be user-specified) on the appropriate phase, and issue an alarm. For a detector that influences the adaptive operation (e.g., a system detector), the system will use data from an alternate (user-specified) detector, such as in an adjacent lane or at an appropriate upstream or downstream location. If the number of detector failures within a specified group exceeds a user-specified threshold, the system will cease adaptive operation and go to a fallback mode of time-of-day operation or free operation. The fallback mode will be specified by the user based on location and time of day.

All detector failure alarms will be automatically transmitted to maintenance and operations staff for appropriate attention.

COMMUNICATION FAILURE

Depending on the architecture of the system, communications failures will have varying effects on the operation. If a communication failure prevents the adaptive system from continuing to control one or more intersections within a defined group, all signals within the group will revert to an appropriate, user-specified fallback mode of operation, either time-of-day operation or free operation. The fallback mode will be specified by the user based on location and time of day.

All communication failure alarms will be automatically transmitted to maintenance and operations staff for appropriate attention.

ADAPTIVE SYSTEM FAILURE

There are two types of adaptive system failures: failure of the server or equipment that operates the adaptive algorithms; and inability of the adaptive algorithms to accommodate current traffic conditions.

If the equipment that operates the adaptive algorithms fails, the system will recognize the failure and place the operation in an appropriate, user-specified fallback mode, either time-of-day operation or free operation. The fallback mode will be specified by the user based and time of day.

The adaptive system makes its decisions based largely on detector information. Occasionally, as the result of an incident or other event outside the control of the system and outside the area covered by the system, congestion will propagate back into the adaptive control area and the measured traffic conditions will be outside the range of data that can be processed by the system. In locations where this is likely to occur, the intersection detectors, or queue detectors installed specifically for this purpose, will measure increased occupancy. In such cases, when user-specified signal timing and detector occupancy conditions are met, the system will recognize that the its response to the input data may not be appropriate, and it will revert to an appropriate, user-specified fallback mode, either time-of-day operation or free operation. The fallback mode will be specified by the user based and time of day.

All adaptive system failure alarms will be automatically and immediately transmitted to maintenance and operations staff for appropriate attention.

ULTIMATE DEPLOYMENT
This section describes several additional operational scenarios that will not occur on the S. Broadway Avenue corridor, but are situations that will need to be covered by the adaptive system elsewhere in Central City.

ISOLATED CRITICAL INTERSECTION
At an isolated intersection that is not part of a coordinated arterial, but cannot operate efficiently as a vehicle-actuated intersection in free mode at all times, the system will determine the appropriate cycle length, phase sequence and split times. User-specified limits will be applied to key parameters such as cycle length and duration between successive displays of a phase. Upon detection of phase failure (through high occupancy of detection zones) or overflow of storage bays (through queue detection), the system will introduce the appropriate phase twice per cycle until the condition is cleared. The system will also have the ability to select different phase sequences to efficiently accommodate different balances of traffic volumes.

CROSSING ARTERIALS
Where crossing arterials are coordinated by the system, the system will determine the appropriate cycle length and direction of coordination independently on each arterial. If the critical intersection requires significantly different operation than the other arterial intersections on all approaches, the system will operate this intersection in the manner described in section 0 for an isolated critical intersection. If the desired operation on both arterials requires a similar cycle length, then both arterials will be coordinated in a consistent manner. Although all intersections may operate at the same cycle length, the system will still independently select the mode and direction of coordination to suit each arterial (e.g., maximum throughput on one arterial and balanced coordination bands on the crossing arterial, even though they are operating with a common cycle length).

If the operation on one arterial requires a different cycle length or other characteristics than are appropriate for the crossing arterial, the intersections on the crossing arterial will operate with a separate coordination pattern and the crossing arterial approaches to the critical intersection will not be coordinated.

CROSS-JURISDICTION OPERATION
At freeway interchanges on some arterials, the signals are owned and/or operated by State DOT. In the future, if these signals are included in the adaptive system, State DOT operators will be able to log on to the adaptive system to monitor and control those intersections, and monitor other elements of the system. This will allow them to set the parameters under which these signals will be coordinated with other (Central City) signals on the arterials, and override the adaptive operation at those intersections

should conditions require it (such as presence of an incident and unacceptable queuing on the freeway off-ramps).

CLOSELY SPACED INTERSECTIONS
For successful operation at the interchanges and other locations with closely spaced intersections, the system will operate within user-specified cycle length and coordination limits that are appropriate for closely spaced intersections. This will include: control of maximum cycle lengths to prevent excessive queuing on internal approaches between intersections; use of trailing offsets to clear blocks at the end of certain phases; and the ability to operate one intersection as a slave to another, so its phase endings clear the block between them, and phase start can be held until the block ahead is guaranteed to be clear.

Appendix D
System Requirements
Table of Sample Requirements

Requirements Document Reference Number	System Requirements Sample Requirements	Need Statement (Con Ops)
1	**1 Network Characteristics**	
1.0-1	The ASCT shall control a minimum of XX signals concurrently	4.2.0-1 The system operator needs to eventually adaptively control up to XXX signals, up to XXX miles from the TMC (or specified location).
1.0-2	The ASCT shall support groups of signals.	4.2.0-2 The system operator needs to be able to adaptively control up to XX independent groups of signals 4.2.0-3 The system operator needs to vary the number of signals in an adaptively controlled group to accommodate the prevailing traffic conditions.
1.0-2.0-1	The boundaries surrounding signal controllers that operate in a coordinated fashion shall be defined by the user.	4.2.0-2 The system operator needs to be able to adaptively control up to XX independent groups of signals
1.0-2.0-2	The ASCT shall control a minimum of XX groups of signals.	4.2.0-2 The system operator needs to be able to adaptively control up to XX independent groups of signals
1.0-2.0-3	The size of a group shall range from 1 to XX signals.	4.2.0-3 The system operator needs to vary the number of signals in an adaptively controlled group to accommodate the prevailing traffic conditions.
1.0-2.0-4	Each group shall operate independently	4.2.0-2 The system operator needs to be able to adaptively control up to XX independent groups of signals
1.0-2.0-5	The boundaries surrounding signal controllers that operate in a coordinated fashion shall be altered by the ASCT system according to configured parameters.	4.2.0-3 The system operator needs to vary the number of signals in an adaptively controlled group to accommodate the prevailing traffic conditions.

Requirements Document Reference Number	System Requirements Sample Requirements	Need Statement (Con Ops)
1.02.0-5.0-1	The boundaries surrounding signal controllers that operate in a coordinated fashion shall be altered by the system according to a time of day schedule. (For example: this may be achieved by assigning signals to different groups or by combining groups.)	4.2.0-3 The system operator needs to vary the number of signals in an adaptively controlled group to accommodate the prevailing traffic conditions.
1.02.0-5.0-2	The boundaries surrounding signal controllers that operate in a coordinated fashion shall be altered by the system according to traffic conditions. (For example: this may be achieved by assigning signals to different groups or by combining groups.)	4.2.0-3 The system operator needs to vary the number of signals in an adaptively controlled group to accommodate the prevailing traffic conditions.
1.02.0-5.0-3	The boundaries surrounding signal controllers that operate in a coordinated fashion shall be altered by the system when commanded by the user.	4.2.0-3 The system operator needs to vary the number of signals in an adaptively controlled group to accommodate the prevailing traffic conditions.
2	**2 Type of Operation**	
2.1	2.1 General	
2.1.1	2.1.1 Mode of Operation	
2.1.1.0-1	The ASCT shall operate non-adaptively during the presence of a defined condition.	4.7.0-1 The system operator needs to detect traffic conditions during which adaptive control is not the preferred operation, and implement some pre-defined operation while that condition is present.
2.1.1.0-2	The ASCT shall operate non-adaptively when adaptive control equipment fails.	4.14.0-1 The system operator needs to fall back to TOD or isolated free operation, as specified by the operator, without causing disruption to traffic flow, in the event of equipment, communications and software failure.
2.1.1.02.0-1	The ASCT shall operate non-adaptively when a user-specified detector fails.	4.14.0-1 The system operator needs to fall back to TOD or isolated free operation, as specified by the operator, without causing disruption to traffic flow, in the event of equipment, communications and software failure.

Requirements Document Reference Number	System Requirements Sample Requirements	Need Statement (Con Ops)
2.1.1.0-2.0-2	The ASCT shall operate non-adaptively when the number of failed detectors connected to a signal controller exceeds a user-defined value.	4.14.0-1 The system operator needs to fall back to TOD or isolated free operation, as specified by the operator, without causing disruption to traffic flow, in the event of equipment, communications and software failure.
2.1.1.0-2.0-3	The ASCT shall operate non-adaptively when the number of failed detectors in a group exceeds a user-defined value.	4.14.0-1 The system operator needs to fall back to TOD or isolated free operation, as specified by the operator, without causing disruption to traffic flow, in the event of equipment, communications and software failure.
2.1.1.0-2.0-4	The ASCT shall operate non-adaptively when a user-defined communications link fails.	4.14.0-1 The system operator needs to fall back to TOD or isolated free operation, as specified by the operator, without causing disruption to traffic flow, in the event of equipment, communications and software failure.
2.1.1.0-3	The ASCT shall operate non-adaptively when a user manually commands the ASCT to cease adaptively controlling a group of signals.	4.7.0-3 The system operator needs to over-ride adaptive operation.
2.1.1.0-4	The ASCT shall operate non-adaptively when a user manually commands the ASCT to cease adaptive operation.	4.7.0-3 The system operator needs to over-ride adaptive operation.
2.1.1.0-5	The ASCT shall operate non-adaptively in accordance with a user-defined time-of-day schedule.	4.7.0-2 The system operator needs to schedule pre-determined operation by time of day. 4.7.0-3 The system operator needs to over-ride adaptive operation
2.1.1.0-6	The ASCT shall operate non-adaptively when commanded by an external system process.	4.17.0-2 The system operator needs to react to commands issued by (specify an external control or decision support system, such as an ICM system or another signal system).

Requirements Document Reference Number	System Requirements Sample Requirements	Need Statement (Con Ops)
2.1.1.0.7	The ASCT shall alter the adaptive operation to achieve required objectives in user-specified conditions. (The required objectives are specified in Needs Statement 4.1.0-1. Responding to this requirement demonstrates how the proposed system allows the user to define the conditions at which the objectives shift and their associated requirements are fulfilled.) (The alteration may be made by adjusting parameters or by directly controlling the state of signal controllers.)	4.1.0-1.0-1 Maximize the throughput on coordinated routes Note to user when selecting these requirements: Select from requirements in the 2.2 group when sequence-based systems are allowed (sequence-based systems explicitly calculate cycle, offset, and split). Select from requirements in the 2.3 group when non-sequence-based systems are allowed (non-sequence-based systems do not explicitly calculate cycle, offset, and split). (Select requirements from both groups when the vendor is given the choice of supplying one type of adaptive operation or the other.) 4.1.0-1.0-3 Distribute phase times in an equitable fashion Note to user when selecting these requirements: Select from requirements in the 2.2 group when sequence-based systems are allowed (sequence-based systems explicitly calculate cycle, offset, and split). Select from requirements in the 2.3 group when non-sequence-based systems are allowed (non-sequence-based systems do not explicitly calculate cycle, offset, and split). (Select requirements from both groups when the vendor is given the choice of supplying one type of adaptive operation or the other.) 4.1.0-3 The system operator needs to change the operational strategy (for example, from smooth flow to maximizing throughput or managing queues) based on changing traffic conditions.

Requirements Document Reference Number	System Requirements Sample Requirements	Need Statement (Con Ops)
2.1.1.0.7.0-1	When current measured traffic conditions meet user-specified criteria, the ASCT shall alter the state of the signal controllers, maximizing the throughput of the coordinated route.	4.1.0-1.0-1 Maximize the throughput on coordinated routes *Note to user when selecting these requirements:* *Select from requirements in the 2.2 group when sequence-based systems are allowed (sequence-based systems explicitly calculate cycle, offset, and split).* *Select from requirements in the 2.3 group when non-sequence-based systems are allowed (non-sequence-based systems do not explicitly calculate cycle, offset, and split).* *(Select requirements from both groups when the vendor is given the choice of supplying one type of adaptive operation or the other.)* 4.1.0-3 The system operator needs to change the operational strategy (for example, from smooth flow to maximizing throughput or managing queues) based on changing traffic conditions.
2.1.1.0.7.0-2	When current measured traffic conditions meet user-specified criteria, the ASCT shall alter the state of signal controllers, preventing queues from exceeding the storage capacity at user-specified locations.	4.1.0-1.0-4 Manage the lengths of queues *Note to user when selecting these requirements:* *Select from requirements in the 2.2 group when sequence-based systems are allowed (sequence-based systems explicitly calculate cycle, offset, and split).* *Select from requirements in the 2.3 group when non-sequence-based systems are allowed (non-sequence-based systems do not explicitly calculate cycle, offset, and split).* *(Select requirements from both groups when the vendor is given the choice of supplying one type of adaptive operation or the other.)*

Requirements Document Reference Number	System Requirements Sample Requirements	Need Statement (Con Ops)
		4.1.0-3 The system operator needs to change the operational strategy (for example, from smooth flow to maximizing throughput or managing queues) based on changing traffic conditions.
2.1.1.0-7.0-3	When current measured traffic conditions meet user-specified criteria, the ASCT shall alter the state of signal controllers providing equitable distribution of green times.	4.1.0-1.0-3 Distribute phase times in an equitable fashion *Note to user when selecting these requirements:* *Select from requirements in the 2.2 group when sequence-based systems are allowed (sequence-based systems explicitly calculate cycle, offset, and split).* *Select from requirements in the 2.3 group when non-sequence-based systems are allowed (non-sequence-based systems do not explicitly calculate cycle, offset, and split).* *(Select requirements from both groups when the vendor is given the choice of supplying one type of adaptive operation or the other.)* 4.1.0-3 The system operator needs to change the operational strategy (for example, from smooth flow to maximizing throughput or managing queues) based on changing traffic conditions.

Requirements Document Reference Number	System Requirements Sample Requirements	Need Statement (Con Ops)
2.1.1.0.7.0.4	When current measured traffic conditions meet user-defined criteria, the ASCT shall alter the state of signal controllers providing two-way progression on a coordinated route.	4.1.0-1.0-2 Provide smooth flow along coordinated routes *Note to user when selecting these requirements:* *Select from requirements in the 2.2 group when sequence-based systems are allowed (sequence-based systems explicitly calculate cycle, offset, and split).* *Select from requirements in the 2.3 group when non-sequence-based systems are allowed (non-sequence-based systems do not explicitly calculate cycle, offset, and split).* *(Select requirements from both groups when the vendor is given the choice of supplying one type of adaptive operation or the other.)* 4.1.0-3 The system operator needs to change the operational strategy (for example, from smooth flow to maximizing throughput or managing queues) based on changing traffic conditions.
2.1.1.0-8	The ASCT shall provide maximum and minimum phase times.	4.1.0-1.0-3 Distribute phase times in an equitable fashion *Note to user when selecting these requirements:* *Select from requirements in the 2.2 group when sequence-based systems are allowed (sequence-based systems explicitly calculate cycle, offset, and split).* *Select from requirements in the 2.3 group when non-sequence-based systems are allowed (non-sequence-based systems do not explicitly calculate cycle, offset, and split).* *(Select requirements from both groups when the vendor is given the choice of supplying one type of adaptive operation or the other.)*

Requirements Document Reference Number	System Requirements Sample Requirements	Need Statement (Con Ops)
		4.1.0-1.0-6 At an isolated intersection, optimize operation with a minimum of phase failures (based on the optimization objectives).
		4.1.0-1.0-3 Distribute phase times in an equitable fashion
		Note to user when selecting these requirements:
		Select from requirements in the 2.2 group when sequence-based systems are allowed (sequence-based systems explicitly calculate cycle, offset, and split).
		Select from requirements in the 2.3 group when non-sequence-based systems are allowed (non-sequence-based systems do not explicitly calculate cycle, offset, and split).
		[Select requirements from both groups when the vendor is given the choice of supplying one type of adaptive operation or the other.]
		4.1.0-1.0-6 At an isolated intersection, optimize operation with a minimum of phase failures (based on the optimization objectives).
2.1.1.0.8.0-1	The ASCT shall provide a user-specified maximum value for each phase at each signal controller.	

Requirements Document Reference Number	System Requirements Sample Requirements	Need Statement (Con Ops)
2.1.1.0.8.0.1.0-1	The ASCT shall not provide a phase length longer that the maximum value.	4.1.0-1.0-3 Distribute phase times in an equitable fashion *Note to user when selecting these requirements:* *Select from requirements in the 2.2 group when sequence-based systems are allowed (sequence-based systems explicitly calculate cycle, offset, and split).* *Select from requirements in the 2.3 group when non-sequence-based systems are allowed (non-sequence-based systems do not explicitly calculate cycle, offset, and split).* *(Select requirements from both groups when the vendor is given the choice of supplying one type of adaptive operation or the other.)* 4.1.0-1.0-6 At an isolated intersection, optimize operation with a minimum of phase failures (based on the optimization objectives).
2.1.1.0.8.0.2	The ASCT shall provide a user-specified minimum value for each phase at each signal controller.	4.1.0-1.0-3 Distribute phase times in an equitable fashion *Note to user when selecting these requirements:* *Select from requirements in the 2.2 group when sequence-based systems are allowed (sequence-based systems explicitly calculate cycle, offset, and split).* *Select from requirements in the 2.3 group when non-sequence-based systems are allowed (non-sequence-based systems do not explicitly calculate cycle, offset, and split).* *(Select requirements from both groups when the vendor is given the choice of supplying one type of adaptive operation or the other.)* 4.1.0-1.0-6 At an isolated intersection, optimize operation with a minimum of phase failures (based on the optimization objectives).

Requirements Document Reference Number	System Requirements Sample Requirements	Need Statement (Con Ops)
2.1.1.0.8.0.2.0-1	The ASCT shall not provide a phase length shorter than the minimum value.	4.1.0-1.0-3 Distribute phase times in an equitable fashion *Note to user when selecting these requirements:* *Select from requirements in the 2.2 group when sequence-based systems are allowed (sequence-based systems explicitly calculate cycle, offset, and split).* *Select from requirements in the 2.3 group when non-sequence-based systems are allowed (non-sequence-based systems do not explicitly calculate cycle, offset, and split).* *(Select requirements from both groups when the vendor is given the choice of supplying one type of adaptive operation or the other.)* 4.1.0-1.0-6 At an isolated intersection, optimize operation with a minimum of phase failures (based on the optimization objectives).
2.1.1.0-9	The ASCT shall detect repeated phases that do not serve all waiting vehicles. (These phase failures may be inferred, such as by detecting repeated max-out.)	4.1.0-4 The system operator needs to detect repeated phase failures and control signal timing to prevent phase failures building up queues. The operator in this case is trying to prevent a routine queue from forming where it will block another movement in the cycle unnecessarily. For example, the operator may need to prevent a queue resulting from the trailing end of the through green from blocking the storage needed by an entering side-street left turn in the subsequent phase. An overall queue management strategy, particularly when congestion is present, is covered under 4.1.0-1.0-5.

Requirements Document Reference Number	System Requirements Sample Requirements	Need Statement (Con Ops)
2.1.1.0-9.0-1	The ASCT shall alter operations, to minimize repeated phase failures.	4.1.0-4 The system operator needs to detect repeated phase failures and control signal timing to prevent phase failures building up queues. The operator in this case is trying to prevent a routine queue from forming where it will block another movement in the cycle unnecessarily. For example, the operator may need to prevent a queue resulting from the trailing end of the through green from blocking the storage needed by an entering side-street left turn in the subsequent phase. An overall queue management strategy, particularly when congestion is present, is covered under 4.1.0-1.0-5.
2.1.1.0-10	The ASCT shall determine the order of phases at a user-specified intersection. (The calculation will be based on the optimization function.)	4.1.0-1.0-1 Maximize the throughput on coordinated routes Note to user when selecting these requirements: Select from requirements in the 2.2 group when sequence-based systems are allowed (sequence-based systems explicitly calculate cycle, offset, and split). Select from requirements in the 2.3 group when non-sequence-based systems are allowed (non-sequence-based systems do not explicitly calculate cycle, offset, and split). (Select requirements from both groups when the vendor is given the choice of supplying one type of adaptive operation or the other.)

Requirements Document Reference Number	System Requirements Sample Requirements	Need Statement (Con Ops)
		4.1.0-1.0-2 Provide smooth flow along coordinated routes *Note to user when selecting these requirements:* *Select from requirements in the 2.2 group when sequence-based systems are allowed (sequence-based systems explicitly calculate cycle, offset, and split).* *Select from requirements in the 2.3 group when non-sequence-based systems are allowed (non-sequence-based systems do not explicitly calculate cycle, offset, and split).* *(Select requirements from both groups when the vendor is given the choice of supplying one type of adaptive operation or the other.)* **4.1.0-1.0-4** Manage the lengths of queues *Note to user when selecting these requirements:* *Select from requirements in the 2.2 group when sequence-based systems are allowed (sequence-based systems explicitly calculate cycle, offset, and split).* *Select from requirements in the 2.3 group when non-sequence-based systems are allowed (non-sequence-based systems do not explicitly calculate cycle, offset, and split).* *(Select requirements from both groups when the vendor is given the choice of supplying one type of adaptive operation or the other.)* **4.1.0-8** The system operator needs to designate the coordinated route based on traffic conditions and the selected operational strategy.
2.1.1.0-11	The ASCT shall provide coordination along a route.	

Requirements Document Reference Number	System Requirements Sample Requirements	Need Statement (Con Ops)
2.1.1.0-11.0-1	The ASCT shall coordinate along a user-defined route.	4.1.0-8 The system operator needs to designate the coordinated route based on traffic conditions and the selected operational strategy.
2.1.1.0-11.0-2	The ASCT shall determine the coordinated route based on traffic conditions.	4.1.0-8 The system operator needs to designate the coordinated route based on traffic conditions and the selected operational strategy.
2.1.1.0-11.0-3	The ASCT shall determine the coordinated route based on a user-defined schedule.	4.1.0-8 The system operator needs to designate the coordinated route based on traffic conditions and the selected operational strategy.
2.1.1.0-11.0-4	The ASCT shall store XX user-defined coordination routes.	4.1.0-8 The system operator needs to designate the coordinated route based on traffic conditions and the selected operational strategy.
2.1.1.0-11.0-4.0-1	The ASCT shall implement a stored coordinated route by operator command.	4.1.0-8 The system operator needs to designate the coordinated route based on traffic conditions and the selected operational strategy.
2.1.1.0-11.0-4.0-2	The ASCT shall implement a stored coordinated route based on traffic conditions.	4.1.0-8 The system operator needs to designate the coordinated route based on traffic conditions and the selected operational strategy.
2.1.1.0-11.0-4.0-3	The ASCT shall implement a stored coordinated route based on a user-defined schedule.	4.1.0-8 The system operator needs to designate the coordinated route based on traffic conditions and the selected operational strategy.
2.1.1.0-12	The ASCT shall not prevent the use of phase timings in the local controller set by agency policy.	4.1.0-9 The system operator needs to set signal timing parameters (such as minimum green, maximum green and extension time) to comply with agency policies.
2.1.2	2.1.2 Allowable Phases	

Requirements Document Reference Number	System Requirements Sample Requirements	Need Statement (Con Ops)
2.1.2.0-1	The ASCT shall not prevent protected/permissive left turn phase operation.	4.9.0-1.0-14 Protected/permissive phasing and alternate left turn phase sequences.
2.1.2.0-2	The ASCT shall not prevent the protected left turn phase to lead or lag the opposing through phase based upon user-specified conditions.	4.9.0-1.0-14 Protected/permissive phasing and alternate left turn phase sequences.
2.1.2.0-3	The ASCT shall prevent skipping a user-specified phase when the user-specified phase sequence is operating.	4.9.0-1.0-6 Prevent one or more phases being skipped under certain traffic conditions or signal states.
2.1.2.0-4	The ASCT shall prevent skipping a user-specified phase based on the state of a user-specified external input.	4.9.0-1.0-6 Prevent one or more phases being skipped under certain traffic conditions or signal states. 4.17.0-2 The system operator needs to react to commands issued by (specify an external control or decision support system, such as an ICM system or another signal system).
2.1.2.0-5	The ASCT shall prevent skipping a user-specified phase according to a time of day schedule.	4.9.0-1.0-6 Prevent one or more phases being skipped under certain traffic conditions or signal states.
2.1.2.0-6	The ASCT shall omit a user-specified phase when the cycle length is below a user-specified value.	4.9.0-1.0-5 Allow one or more phases to be omitted (disabled) under certain traffic conditions or signal states.
2.1.2.0-7	The ASCT shall omit a user-specified phase based on measured traffic conditions.	4.9.0-1.0-5 Allow one or more phases to be omitted (disabled) under certain traffic conditions or signal states. 4.17.0-2 The system operator needs to react to commands issued by (specify an external control or decision support system, such as an ICM system or another signal system).
2.1.2.0-9	The ASCT shall omit a user-specified phase according to a time of day schedule	4.9.0-1.0-5 Allow one or more phases to be omitted (disabled) under certain traffic conditions or signal states.

Requirements Document Reference Number	System Requirements Sample Requirements	Need Statement (Con Ops)
2.1.2.0-10	The ASCT shall assign unused time from a preceding phase that terminates early to a user-specified phase as follows: • next phase • next coordinated phase • user-specified phase	4.9.0-1.0-10 Allow the operator to specify which phase receives unused time from a preceding phase
2.1.2.0-11	The ASCT shall assign unused time from a preceding phase that is skipped to a user-specified phase as follows: • previous phase • next phase • next coordinated phase • user-specified phase	4.9.0-1.0-10 Allow the operator to specify which phase receives unused time from a preceding phase
2.1.2.0-12	The ASCT shall not alter the order of phases at a user-specified intersection.	4.1.0-7 The system operator needs to fix the sequence of phases at any specified location. For example, the operator may need to fix the phase order at a diamond interchange.
2.1.3	2.1.3 Oversaturation	
2.1.3.0-1	The ASCT shall detect the presence of queues at pre-configured locations.	4.1.0-1.0-4 Manage the lengths of queues Note to user when selecting these requirements: *Select from requirements in the 2.2 group when sequence-based systems are allowed (sequence-based systems explicitly calculate cycle, offset, and split).* *Select from requirements in the 2.3 group when non-sequence-based systems are allowed (non-sequence-based systems do not explicitly calculate cycle, offset, and split).* *(Select requirements from both groups when the vendor is given the choice of supplying one type of adaptive operation or the other.)*

Requirements Document Reference Number	System Requirements Sample Requirements	Need Statement (Con Ops)
		4.1.0-1.0-5 Manage the locations of queues within the network *Note to user when selecting these requirements:* *Select from requirements in the 2.2 group when sequence-based systems are allowed (sequence-based systems explicitly calculate cycle, offset, and split).* *Select from requirements in the 2.3 group when non-sequence-based systems are allowed (non-sequence-based systems do not explicitly calculate cycle, offset, and split).* *(Select requirements from both groups when the vendor is given the choice of supplying one type of adaptive operation or the other.)* **4.1.0-4** The system operator needs to detect repeated phase failures and control signal timing to prevent phase failures building up queues. The operator in this case is trying to prevent a routine queue from forming where it will block another movement in the cycle unnecessarily. For example, the operator may need to prevent a queue resulting from the trailing end of the through green from blocking the storage needed by an entering side-street left turn in the subsequent phase. An overall queue management strategy, particularly when congestion is present, is covered under 4.1.0-1.0-5. **4.5.0-1** The system operator needs to detect queues from outside the system and modify the ASCT operation to accommodate the queuing.

Model Systems Engineering Documents for Adaptive Signal Control Technology (ASCT) Systems

Requirements Document Reference Number	System Requirements Sample Requirements	Need Statement (Con Ops)
		4.5.0-2 The system operator needs to detect queues within the system's boundaries and modify the ASCT operation to accommodate the queuing.
		4.5.0-3 The system operator needs to detect queues propagating outside its boundaries from within the ASCT boundaries, and modify its operation to accommodate the queuing.
		4.5.0-4 The system operator needs to store queues in locations where they can be accommodated without adversely affecting adaptive operation.
		4.5.0-5 The system operator needs to prevent queues forming at user-specified locations.
		4.1.0-1.0-4 Manage the lengths of queues
		Note to user when selecting these requirements:
		Select from requirements in the 2.2 group when sequence-based systems are allowed (sequence-based systems explicitly calculate cycle, offset, and split).
		Select from requirements in the 2.3 group when non-sequence-based systems are allowed (non-sequence-based systems do not explicitly calculate cycle, offset, and split).
		(Select requirements from both groups when the vendor is given the choice of supplying one type of adaptive operation or the other.)
2.1.3.0-2	When queues are detected at user-specified locations, the ASCT shall execute user-specified timing plan/operational mode.	

Requirements Document Reference Number	System Requirements Sample Requirements	Need Statement (Con Ops)
		4.1.0-1.0-5 Manage the locations of queues within the network *Note to user when selecting these requirements:* *Select from requirements in the 2.2 group when sequence-based systems are allowed (sequence-based systems explicitly calculate cycle, offset, and split).* *Select from requirements in the 2.3 group when non-sequence-based systems are allowed (non-sequence-based systems do not explicitly calculate cycle, offset, and split).* *(Select requirements from both groups when the vendor is given the choice of supplying one type of adaptive operation or the other.)* **4.1.0-4** The system operator needs to detect repeated phase failures and control signal timing to prevent phase failures building up queues. The operator in this case is trying to prevent a routine queue from forming where it will block another movement in the cycle unnecessarily. For example, the operator may need to prevent a queue resulting from the trailing end of the through green from blocking the storage needed by an entering side-street left turn in the subsequent phase. An overall queue management strategy, particularly when congestion is present, is covered under 4.1.0-1.0-5. **4.5.0-1** The system operator needs to detect queues from outside the system and modify the ASCT operation to accommodate the queuing.

Requirements Document Reference Number	System Requirements Sample Requirements	Need Statement (Con Ops)
		4.5.0-2 The system operator needs to detect queues within the system's boundaries and modify the ASCT operation to accommodate the queuing. 4.5.0-3 The system operator needs to detect queues propagating outside its boundaries from within the ASCT boundaries, and modify its operation to accommodate the queuing. 4.5.0-4 The system operator needs to store queues in locations where they can be accommodated without adversely affecting adaptive operation. 4.5.0-5 The system operator needs to prevent queues forming at user-specified locations.
2.1.3.0-3	When queues are detected at user-specified locations, the ASCT shall execute user-specified adaptive operation strategy.	4.1.0-1.0-4 Manage the lengths of queues *Note to user when selecting these requirements:* *Select from requirements in the 2.2 group when sequence-based systems are allowed (sequence-based systems explicitly calculate cycle, offset, and split).* *Select from requirements in the 2.3 group when non-sequence-based systems are allowed (non-sequence-based systems do not explicitly calculate cycle, offset, and split).* *(Select requirements from both groups when the vendor is given the choice of supplying one type of adaptive operation or the other.)*

Requirements Document Reference Number	System Requirements Sample Requirements	Need Statement (Con Ops)
		4.1.0-1.0-5 Manage the locations of queues within the network *Note to user when selecting these requirements:* *Select from requirements in the 2.2 group when sequence-based systems are allowed (sequence-based systems explicitly calculate cycle, offset, and split).* *Select from requirements in the 2.3 group when non-sequence-based systems are allowed (non-sequence-based systems do not explicitly calculate cycle, offset, and split).* *(Select requirements from both groups when the vendor is given the choice of supplying one type of adaptive operation or the other.)* **4.1.0-4** The system operator needs to detect repeated phase failures and control signal timing to prevent phase failures building up queues. The operator in this case is trying to prevent a routine queue from forming where it will block another movement in the cycle unnecessarily. For example, the operator may need to prevent a queue resulting from the trailing end of the through green from blocking the storage needed by an entering side-street left turn in the subsequent phase. An overall queue management strategy, particularly when congestion is present, is covered under 4.1.0-1.0-5. **4.5.0-1** The system operator needs to detect queues from outside the system and modify the ASCT operation to accommodate the queuing.

Requirements Document Reference Number	System Requirements Sample Requirements	Need Statement (Con Ops)
		4.5.0-2 The system operator needs to detect queues within the system's boundaries and modify the ASCT operation to accommodate the queuing.
		4.5.0-3 The system operator needs to detect queues propagating outside its boundaries from within the ASCT boundaries, and modify its operation to accommodate the queuing.
		4.5.0-4 The system operator needs to store queues in locations where they can be accommodated without adversely affecting adaptive operation.
		4.5.0-5 The system operator needs to prevent queues forming at user-specified locations.
2.1.3.0-4	When queues are detected at user-specified locations, the ASCT shall omit a user-specified phase at a user-specified signal controller.	**4.1.0-1.0-4** Manage the lengths of queues *Note to user when selecting these requirements:* *Select from requirements in the 2.2 group when sequence-based systems are allowed (sequence-based systems explicitly calculate cycle, offset, and split).* *Select from requirements in the 2.3 group when non-sequence-based systems are allowed (non-sequence-based systems do not explicitly calculate cycle, offset, and split).* *(Select requirements from both groups when the vendor is given the choice of supplying one type of adaptive operation or the other.)*

196 — Model Systems Engineering Documents for Adaptive Signal Control Technology (ASCT) Systems

Requirements Document Reference Number	System Requirements Sample Requirements	Need Statement (Con Ops)
		4.1.0-1.0-5 Manage the locations of queues within the network *Note to user when selecting these requirements:* *Select from requirements in the 2.2 group when sequence-based systems are allowed (sequence-based systems explicitly calculate cycle, offset, and split).* *Select from requirements in the 2.3 group when non-sequence-based systems are allowed (non-sequence-based systems do not explicitly calculate cycle, offset, and split).* *[Select requirements from both groups when the vendor is given the choice of supplying one type of adaptive operation or the other.]* 4.1.0-4 The system operator needs to detect repeated phase failures and control signal timing to prevent phase failures building up queues. The operator in this case is trying to prevent a routine queue from forming where it will block another movement in the cycle unnecessarily. For example, the operator may need to prevent a queue resulting from the trailing end of the through green from blocking the storage needed by an entering side-street left turn in the subsequent phase. An overall queue management strategy, particularly when congestion is present, is covered under 4.1.0-1.0-5. 4.5.0-4 The system operator needs to store queues in locations where they can be accommodated without adversely affecting adaptive operation. 4.5.0-5 The system operator needs to prevent queues forming at user-specified locations.

Requirements Document Reference Number	System Requirements Sample Requirements	Need Statement (Con Ops)
2.1.3.0-5	The ASCT shall meter traffic into user-specified bottlenecks by storing queues at user-specified locations.	4.1.0-1.0-4 Manage the lengths of queues *Note to user when selecting these requirements:* *Select from requirements in the 2.2 group when sequence-based systems are allowed (sequence-based systems explicitly calculate cycle, offset, and split).* *Select from requirements in the 2.3 group when non-sequence-based systems are allowed (non-sequence-based systems do not explicitly calculate cycle, offset, and split).* *(Select requirements from both groups when the vendor is given the choice of supplying one type of adaptive operation or the other.)* 4.1.0-1.0-5 Manage the locations of queues within the network *Note to user when selecting these requirements:* *Select from requirements in the 2.2 group when sequence-based systems are allowed (sequence-based systems explicitly calculate cycle, offset, and split).* *Select from requirements in the 2.3 group when non-sequence-based systems are allowed (non-sequence-based systems do not explicitly calculate cycle, offset, and split).* *(Select requirements from both groups when the vendor is given the choice of supplying one type of adaptive operation or the other.)* 4.5.0-4 The system operator needs to store queues in locations where they can be accommodated without adversely affecting adaptive operation.

Requirements Document Reference Number	System Requirements Sample Requirements	Need Statement (Con Ops)
		4.5.0-5 The system operator needs to prevent queues forming at user-specified locations.
2.1.3.0-6	The ASCT shall store queues at user-specified locations.	4.1.0-1.0-4 Manage the lengths of queues *Note to user when selecting these requirements:* *Select from requirements in the 2.2 group when sequence-based systems are allowed (sequence-based systems explicitly calculate cycle, offset, and split).* *Select from requirements in the 2.3 group when non-sequence-based systems are allowed (non-sequence-based systems do not explicitly calculate cycle, offset, and split).* *(Select requirements from both groups when the vendor is given the choice of supplying one type of adaptive operation or the other.)* 4.1.0-1.0-5 Manage the locations of queues within the network *Note to user when selecting these requirements:* *Select from requirements in the 2.2 group when sequence-based systems are allowed (sequence-based systems explicitly calculate cycle, offset, and split).* *Select from requirements in the 2.3 group when non-sequence-based systems are allowed (non-sequence-based systems do not explicitly calculate cycle, offset, and split).* *(Select requirements from both groups when the vendor is given the choice of supplying one type of adaptive operation or the other.)*

Requirements Document Reference Number	System Requirements Sample Requirements	Need Statement (Con Ops)
		4.5.0-4 The system operator needs to store queues in locations where they can be accommodated without adversely affecting adaptive operation.
		4.5.0-5 The system operator needs to prevent queues forming at user-specified locations.
2.1.3.0-7	The ASCT shall maintain capacity flow through user-specified bottlenecks.	4.5.0-4 The system operator needs to store queues in locations where they can be accommodated without adversely affecting adaptive operation.
		4.5.0-5 The system operator needs to prevent queues forming at user-specified locations.
2.1.3.0-8	When queues are detected at user-specified locations, the ASCT shall limit the cycle length of the group to a user-specified value.	4.1.0-1.0-5 Manage the locations of queues within the network *Note to user when selecting these requirements:* *Select from requirements in the 2.2 group when sequence-based systems are allowed (sequence-based systems explicitly calculate cycle, offset, and split).* *Select from requirements in the 2.3 group when non-sequence-based systems are allowed (non-sequence-based systems do not explicitly calculate cycle, offset, and split).* *(Select requirements from both groups when the vendor is given the choice of supplying one type of adaptive operation or the other.)*
2.2	2.2 Sequence-based Adaptive Coordination	
2.2.0-1	Use this section if sequence-based adaptive coordination is likely to provide acceptable operation in your situation.	

Requirements Document Reference Number	System Requirements Sample Requirements	Need Statement (Con Ops)
2.2.0-2	(Sequence-based only) The ASCT shall select cycle length based on a time of day schedule.	4.1.0-1.0-1 Maximize the throughput on coordinated routes *Note to user when selecting these requirements:* *Select from requirements in the 2.2 group when sequence-based systems are allowed (sequence-based systems explicitly calculate cycle, offset, and split).* *Select from requirements in the 2.3 group when non-sequence-based systems are allowed (non-sequence-based systems do not explicitly calculate cycle, offset, and split).* *(Select requirements from both groups when the vendor is given the choice of supplying one type of adaptive operation or the other.)* 4.1.0-1.0-2 Provide smooth flow along coordinated routes *Note to user when selecting these requirements:* *Select from requirements in the 2.2 group when sequence-based systems are allowed (sequence-based systems explicitly calculate cycle, offset, and split).* *Select from requirements in the 2.3 group when non-sequence-based systems are allowed (non-sequence-based systems do not explicitly calculate cycle, offset, and split).* *(Select requirements from both groups when the vendor is given the choice of supplying one type of adaptive operation or the other.)*

Requirements Document Reference Number	System Requirements Sample Requirements	Need Statement (Con Ops)
		4.1.0-1.0-3 Distribute phase times in an equitable fashion *Note to user when selecting these requirements:* *Select from requirements in the 2.2 group when sequence-based systems are allowed (sequence-based systems explicitly calculate cycle, offset, and split).* *Select from requirements in the 2.3 group when non-sequence-based systems are allowed (non-sequence-based systems do not explicitly calculate cycle, offset, and split).* *(Select requirements from both groups when the vendor is given the choice of supplying one type of adaptive operation or the other.)* 4.1.0-1.0-4 Manage the lengths of queues *Note to user when selecting these requirements:* *Select from requirements in the 2.2 group when sequence-based systems are allowed (sequence-based systems explicitly calculate cycle, offset, and split).* *Select from requirements in the 2.3 group when non-sequence-based systems are allowed (non-sequence-based systems do not explicitly calculate cycle, offset, and split).* *(Select requirements from both groups when the vendor is given the choice of supplying one type of adaptive operation or the other.)*

Requirements Document Reference Number	System Requirements Sample Requirements	Need Statement (Con Ops)
2.2.0.3	(Sequence-based only) The ASCT shall calculate phase lengths for all phases at each signal controller to suit the current coordination strategy.	4.1.0-1.0-3 Distribute phase times in an equitable fashion *Note to user when selecting these requirements:* *Select from requirements in the 2.2 group when sequence-based systems are allowed (sequence-based systems explicitly calculate cycle, offset, and split).* *Select from requirements in the 2.3 group when non-sequence-based systems are allowed (non-sequence-based systems do not explicitly calculate cycle, offset, and split).* *(Select requirements from both groups when the vendor is given the choice of supplying one type of adaptive operation or the other.)* 4.1.0-1.0-5 Manage the locations of queues within the network *Note to user when selecting these requirements:* *Select from requirements in the 2.2 group when sequence-based systems are allowed (sequence-based systems explicitly calculate cycle, offset, and split).* *Select from requirements in the 2.3 group when non-sequence-based systems are allowed (non-sequence-based systems do not explicitly calculate cycle, offset, and split).* *(Select requirements from both groups when the vendor is given the choice of supplying one type of adaptive operation or the other.)*

Requirements Document Reference Number	System Requirements Sample Requirements	Need Statement (Con Ops)
		4.1.0-4 *The system operator needs to detect repeated phase failures and control signal timing to prevent phase failures building up queues. The operator in this case is trying to prevent a routine queue from forming where it will block another movement in the cycle unnecessarily. For example, the operator may need to prevent a queue resulting from the trailing end of the through green from blocking the storage needed by an entering side-street left turn in the subsequent phase. An overall queue management strategy, particularly when congestion is present, is covered under 4.1.0-1.0-5.*
		4.1.0-1.0-1 *Maximize the throughput on coordinated routes* *Note to user when selecting these requirements:* *Select from requirements in the 2.2 group when sequence-based systems are allowed (sequence-based systems explicitly calculate cycle, offset, and split).* *Select from requirements in the 2.3 group when non-sequence-based systems are allowed (non-sequence-based systems do not explicitly calculate cycle, offset, and split).* *[Select requirements from both groups when the vendor is given the choice of supplying one type of adaptive operation or the other.]*
2.2.0-4	(Sequence-based only) The ASCT shall calculate offsets to suit the current coordination strategy for the user-specified reference point for each signal controller along a coordinated route within a group.	

Requirements Document Reference Number	System Requirements Sample Requirements	Need Statement (Con Ops)
		4.1.0-1.0-2 Provide smooth flow along coordinated routes *Note to user when selecting these requirements:* *Select from requirements in the 2.2 group when sequence-based systems are allowed (sequence-based systems explicitly calculate cycle, offset, and split).* *Select from requirements in the 2.3 group when non-sequence-based systems are allowed (non-sequence-based systems do not explicitly calculate cycle, offset, and split).* *(Select requirements from both groups when the vendor is given the choice of supplying one type of adaptive operation or the other.)* 4.1.0-1.0-4 Manage the lengths of queues *Note to user when selecting these requirements:* *Select from requirements in the 2.2 group when sequence-based systems are allowed (sequence-based systems explicitly calculate cycle, offset, and split).* *Select from requirements in the 2.3 group when non-sequence-based systems are allowed (non-sequence-based systems do not explicitly calculate cycle, offset, and split).* *(Select requirements from both groups when the vendor is given the choice of supplying one type of adaptive operation or the other.)*

Requirements Document Reference Number	System Requirements Sample Requirements	Need Statement (Con Ops)
2.2.0.4.0-1	(Sequence-based only) The ASCT shall apply offsets for the user-specified reference point of each signal controller along a coordinated route.	4.1.0-1.0-1 Maximize the throughput on coordinated routes *Note to user when selecting these requirements:* *Select from requirements in the 2.2 group when sequence-based systems are allowed (sequence-based systems explicitly calculate cycle, offset, and split).* *Select from requirements in the 2.3 group when non-sequence-based systems are allowed (non-sequence-based systems do not explicitly calculate cycle, offset, and split).* *(Select requirements from both groups when the vendor is given the choice of supplying one type of adaptive operation or the other.)* 4.1.0-1.0-2 Provide smooth flow along coordinated routes *Note to user when selecting these requirements:* *Select from requirements in the 2.2 group when sequence-based systems are allowed (sequence-based systems explicitly calculate cycle, offset, and split).* *Select from requirements in the 2.3 group when non-sequence-based systems are allowed (non-sequence-based systems do not explicitly calculate cycle, offset, and split).* *(Select requirements from both groups when the vendor is given the choice of supplying one type of adaptive operation or the other.)*

Requirements Document Reference Number	System Requirements Sample Requirements	Need Statement (Con Ops)
		4.1.0-1.0-4 Manage the lengths of queues *Note to user when selecting these requirements:* *Select from requirements in the 2.2 group when sequence-based systems are allowed (sequence-based systems explicitly calculate cycle, offset, and split).* *Select from requirements in the 2.3 group when non-sequence-based systems are allowed (non-sequence-based systems do not explicitly calculate cycle, offset, and split).* *(Select requirements from both groups when the vendor is given the choice of supplying one type of adaptive operation or the other.)*
2.2.0-5	(Sequence-based only) The ASCT shall calculate a cycle length for each cycle based on its optimization objectives (as required elsewhere, e.g., progression, queue management, equitable distribution of green).	4.1.0-1.0-1 Maximize the throughput on coordinated routes *Note to user when selecting these requirements:* *Select from requirements in the 2.2 group when sequence-based systems are allowed (sequence-based systems explicitly calculate cycle, offset, and split).* *Select from requirements in the 2.3 group when non-sequence-based systems are allowed (non-sequence-based systems do not explicitly calculate cycle, offset, and split).* *(Select requirements from both groups when the vendor is given the choice of supplying one type of adaptive operation or the other.)*

Requirements Document Reference Number	System Requirements Sample Requirements	Need Statement (Con Ops)
		4.1.0-1.0-2 Provide smooth flow along coordinated routes *Note to user when selecting these requirements:* *Select from requirements in the 2.2 group when sequence-based systems are allowed (sequence-based systems explicitly calculate cycle, offset, and split).* *Select from requirements in the 2.3 group when non-sequence-based systems are allowed (non-sequence-based systems do not explicitly calculate cycle, offset, and split).* *(Select requirements from both groups when the vendor is given the choice of supplying one type of adaptive operation or the other.)* 4.1.0-1.0-3 Distribute phase times in an equitable fashion *Note to user when selecting these requirements:* *Select from requirements in the 2.2 group when sequence-based systems are allowed (sequence-based systems explicitly calculate cycle, offset, and split).* *Select from requirements in the 2.3 group when non-sequence-based systems are allowed (non-sequence-based systems do not explicitly calculate cycle, offset, and split).* *(Select requirements from both groups when the vendor is given the choice of supplying one type of adaptive operation or the other.)*

Requirements Document Reference Number	System Requirements Sample Requirements	Need Statement (Con Ops)
		4.1.0-1.0-4 Manage the lengths of queues *Note to user when selecting these requirements:* *Select from requirements in the 2.2 group when sequence-based systems are allowed (sequence-based systems explicitly calculate cycle, offset, and split).* *Select from requirements in the 2.3 group when non-sequence-based systems are allowed (non-sequence-based systems do not explicitly calculate cycle, offset, and split).* *(Select requirements from both groups when the vendor is given the choice of supplying one type of adaptive operation or the other.)* **4.1.0-1.0-1** Maximize the throughput on coordinated routes *Note to user when selecting these requirements:* *Select from requirements in the 2.2 group when sequence-based systems are allowed (sequence-based systems explicitly calculate cycle, offset, and split).* *Select from requirements in the 2.3 group when non-sequence-based systems are allowed (non-sequence-based systems do not explicitly calculate cycle, offset, and split).* *(Select requirements from both groups when the vendor is given the choice of supplying one type of adaptive operation or the other.)*
2.2.0-5.0-1	(Sequence-based only) The ASCT shall limit cycle lengths to user-specified values.	

Requirements Document Reference Number	System Requirements Sample Requirements	Need Statement (Con Ops)
		4.1.0-1.0-2 Provide smooth flow along coordinated routes *Note to user when selecting these requirements:* *Select from requirements in the 2.2 group when sequence-based systems are allowed (sequence-based systems explicitly calculate cycle, offset, and split).* *Select from requirements in the 2.3 group when non-sequence-based systems are allowed (non-sequence-based systems do not explicitly calculate cycle, offset, and split).* *(Select requirements from both groups when the vendor is given the choice of supplying one type of adaptive operation or the other.)* 4.1.0-1.0-3 Distribute phase times in an equitable fashion *Note to user when selecting these requirements:* *Select from requirements in the 2.2 group when sequence-based systems are allowed (sequence-based systems explicitly calculate cycle, offset, and split).* *Select from requirements in the 2.3 group when non-sequence-based systems are allowed (non-sequence-based systems do not explicitly calculate cycle, offset, and split).* *(Select requirements from both groups when the vendor is given the choice of supplying one type of adaptive operation or the other.)*

Requirements Document Reference Number	System Requirements Sample Requirements	Need Statement (Con Ops)
		4.1.0-1.0-4 Manage the lengths of queues *Note to user when selecting these requirements:* *Select from requirements in the 2.2 group when sequence-based systems are allowed (sequence-based systems explicitly calculate cycle, offset, and split).* *Select from requirements in the 2.3 group when non-sequence-based systems are allowed (non-sequence-based systems do not explicitly calculate cycle, offset, and split).* *(Select requirements from both groups when the vendor is given the choice of supplying one type of adaptive operation or the other.)*
		4.1.0-1.0-1 Maximize the throughput on coordinated routes *Note to user when selecting these requirements:* *Select from requirements in the 2.2 group when sequence-based systems are allowed (sequence-based systems explicitly calculate cycle, offset, and split).* *Select from requirements in the 2.3 group when non-sequence-based systems are allowed (non-sequence-based systems do not explicitly calculate cycle, offset, and split).* *(Select requirements from both groups when the vendor is given the choice of supplying one type of adaptive operation or the other.)*
2.2.0-5.0-2	(Sequence-based only) The ASCT shall limit cycle lengths to a user-specified range.	

Requirements Document Reference Number	System Requirements Sample Requirements	Need Statement (Con Ops)
		4.1.0-1.0-2 Provide smooth flow along coordinated routes *Note to user when selecting these requirements:* *Select from requirements in the 2.2 group when sequence-based systems are allowed (sequence-based systems explicitly calculate cycle, offset, and split).* *Select from requirements in the 2.3 group when non-sequence-based systems are allowed (non-sequence-based systems do not explicitly calculate cycle, offset, and split).* *(Select requirements from both groups when the vendor is given the choice of supplying one type of adaptive operation or the other.)* 4.1.0-1.0-3 Distribute phase times in an equitable fashion *Note to user when selecting these requirements:* *Select from requirements in the 2.2 group when sequence-based systems are allowed (sequence-based systems explicitly calculate cycle, offset, and split).* *Select from requirements in the 2.3 group when non-sequence-based systems are allowed (non-sequence-based systems do not explicitly calculate cycle, offset, and split).* *(Select requirements from both groups when the vendor is given the choice of supplying one type of adaptive operation or the other.)*

Requirements Document Reference Number	System Requirements Sample Requirements	Need Statement (Con Ops)
		4.1.0-1.0.4 Manage the lengths of queues Note to user when selecting these requirements: Select from requirements in the 2.2 group when sequence-based systems are allowed (sequence-based systems explicitly calculate cycle, offset, and split). Select from requirements in the 2.3 group when non-sequence-based systems are allowed (non-sequence-based systems do not explicitly calculate cycle, offset, and split). (Select requirements from both groups when the vendor is given the choice of supplying one type of adaptive operation or the other.)
2.2.0.5.0.3	(Sequence-based only) The ASCT shall calculate optimum cycle length according to the user-specified coordination strategy.	4.1.0-1.0-1 Maximize the throughput on coordinated routes Note to user when selecting these requirements: Select from requirements in the 2.2 group when sequence-based systems are allowed (sequence-based systems explicitly calculate cycle, offset, and split). Select from requirements in the 2.3 group when non-sequence-based systems are allowed (non-sequence-based systems do not explicitly calculate cycle, offset, and split). (Select requirements from both groups when the vendor is given the choice of supplying one type of adaptive operation or the other.)

Requirements Document Reference Number	System Requirements Sample Requirements	Need Statement (Con Ops)
		4.1.0-1.0-2 Provide smooth flow along coordinated routes *Note to user when selecting these requirements:* *Select from requirements in the 2.2 group when sequence-based systems are allowed (sequence-based systems explicitly calculate cycle, offset, and split).* *Select from requirements in the 2.3 group when non-sequence-based systems are allowed (non-sequence-based systems do not explicitly calculate cycle, offset, and split).* *[Select requirements from both groups when the vendor is given the choice of supplying one type of adaptive operation or the other.]* 4.1.0-1.0-3 Distribute phase times in an equitable fashion *Note to user when selecting these requirements:* *Select from requirements in the 2.2 group when sequence-based systems are allowed (sequence-based systems explicitly calculate cycle, offset, and split).* *Select from requirements in the 2.3 group when non-sequence-based systems are allowed (non-sequence-based systems do not explicitly calculate cycle, offset, and split).* *[Select requirements from both groups when the vendor is given the choice of supplying one type of adaptive operation or the other.]*

Requirements Document Reference Number	System Requirements Sample Requirements	Need Statement (Con Ops)
		4.1.0-1.0-4 Manage the lengths of queues *Note to user when selecting these requirements:* *Select from requirements in the 2.2 group when sequence-based systems are allowed (sequence-based systems explicitly calculate cycle, offset, and split).* *Select from requirements in the 2.3 group when non-sequence-based systems are allowed (non-sequence-based systems do not explicitly calculate cycle, offset, and split).* *(Select requirements from both groups when the vendor is given the choice of supplying one type of adaptive operation or the other.)* 4.1.0-1.0-1 Maximize the throughput on coordinated routes *Note to user when selecting these requirements:* *Select from requirements in the 2.2 group when sequence-based systems are allowed (sequence-based systems explicitly calculate cycle, offset, and split).* *Select from requirements in the 2.3 group when non-sequence-based systems are allowed (non-sequence-based systems do not explicitly calculate cycle, offset, and split).* *(Select requirements from both groups when the vendor is given the choice of supplying one type of adaptive operation or the other.)*
2.2.0-5.0-4	(Sequence-based only) The ASCT shall limit changes in cycle length to not exceed a user-specified value.	

Requirements Document Reference Number	System Requirements Sample Requirements	Need Statement (Con Ops)
		4.1.0-1.0-2 Provide smooth flow along coordinated routes *Note to user when selecting these requirements:* *Select from requirements in the 2.2 group when sequence-based systems are allowed (sequence-based systems explicitly calculate cycle, offset, and split).* *Select from requirements in the 2.3 group when non-sequence-based systems are allowed (non-sequence-based systems do not explicitly calculate cycle, offset, and split).* *(Select requirements from both groups when the vendor is given the choice of supplying one type of adaptive operation or the other.)* 4.1.0-1.0-3 Distribute phase times in an equitable fashion *Note to user when selecting these requirements:* *Select from requirements in the 2.2 group when sequence-based systems are allowed (sequence-based systems explicitly calculate cycle, offset, and split).* *Select from requirements in the 2.3 group when non-sequence-based systems are allowed (non-sequence-based systems do not explicitly calculate cycle, offset, and split).* *(Select requirements from both groups when the vendor is given the choice of supplying one type of adaptive operation or the other.)*

Requirements Document Reference Number	System Requirements Sample Requirements	Need Statement (Con Ops)
		4.1.0-1.0-4 Manage the lengths of queues *Note to user when selecting these requirements:* *Select from requirements in the 2.2 group when sequence-based systems are allowed (sequence-based systems explicitly calculate cycle, offset, and split).* *Select from requirements in the 2.3 group when non-sequence-based systems are allowed (non-sequence-based systems do not explicitly calculate cycle, offset, and split).* *(Select requirements from both groups when the vendor is given the choice of supplying one type of adaptive operation or the other.)*
2.2.0-5.0-4.0-1	(Sequence-based only) The ASCT shall increase the limit for the following XX cycles based on a change in conditions.	**4.1.0-1.0-1** Maximize the throughput on coordinated routes *Note to user when selecting these requirements:* *Select from requirements in the 2.2 group when sequence-based systems are allowed (sequence-based systems explicitly calculate cycle, offset, and split).* *Select from requirements in the 2.3 group when non-sequence-based systems are allowed (non-sequence-based systems do not explicitly calculate cycle, offset, and split).* *(Select requirements from both groups when the vendor is given the choice of supplying one type of adaptive operation or the other.)*

Requirements Document Reference Number	System Requirements Sample Requirements	Need Statement (Con Ops)
		4.1.0-1.0-2 Provide smooth flow along coordinated routes *Note to user when selecting these requirements:* *Select from requirements in the 2.2 group when sequence-based systems are allowed (sequence-based systems explicitly calculate cycle, offset, and split).* *Select from requirements in the 2.3 group when non-sequence-based systems are allowed (non-sequence-based systems do not explicitly calculate cycle, offset, and split).* *(Select requirements from both groups when the vendor is given the choice of supplying one type of adaptive operation or the other.)* 4.1.0-1.0-3 Distribute phase times in an equitable fashion *Note to user when selecting these requirements:* *Select from requirements in the 2.2 group when sequence-based systems are allowed (sequence-based systems explicitly calculate cycle, offset, and split).* *Select from requirements in the 2.3 group when non-sequence-based systems are allowed (non-sequence-based systems do not explicitly calculate cycle, offset, and split).* *(Select requirements from both groups when the vendor is given the choice of supplying one type of adaptive operation or the other.)*

Requirements Document Reference Number	System Requirements Sample Requirements	Need Statement (Con Ops)
		4.1.0-1.0-4 Manage the lengths of queues *Note to user when selecting these requirements:* *Select from requirements in the 2.2 group when sequence-based systems are allowed (sequence-based systems explicitly calculate cycle, offset, and split).* *Select from requirements in the 2.3 group when non-sequence-based systems are allowed (non-sequence-based systems do not explicitly calculate cycle, offset, and split).* *(Select requirements from both groups when the vendor is given the choice of supplying one type of adaptive operation or the other.)*
2.2.0-5.04.0-1.0-1	(Sequence-based only) The change in conditions shall be defined by XX successive adaptive increases in cycle length at the maximum rate.	4.1.0-1.0-1 Maximize the throughput on coordinated routes *Note to user when selecting these requirements:* *Select from requirements in the 2.2 group when sequence-based systems are allowed (sequence-based systems explicitly calculate cycle, offset, and split).* *Select from requirements in the 2.3 group when non-sequence-based systems are allowed (non-sequence-based systems do not explicitly calculate cycle, offset, and split).* *(Select requirements from both groups when the vendor is given the choice of supplying one type of adaptive operation or the other.)*

Requirements Document Reference Number	System Requirements Sample Requirements	Need Statement (Con Ops)
		4.1.0-1.0-2 Provide smooth flow along coordinated routes *Note to user when selecting these requirements:* *Select from requirements in the 2.2 group when sequence-based systems are allowed (sequence-based systems explicitly calculate cycle, offset, and split).* *Select from requirements in the 2.3 group when non-sequence-based systems are allowed (non-sequence-based systems do not explicitly calculate cycle, offset, and split).* *(Select requirements from both groups when the vendor is given the choice of supplying one type of adaptive operation or the other.)* 4.1.0-1.0-3 Distribute phase times in an equitable fashion *Note to user when selecting these requirements:* *Select from requirements in the 2.2 group when sequence-based systems are allowed (sequence-based systems explicitly calculate cycle, offset, and split).* *Select from requirements in the 2.3 group when non-sequence-based systems are allowed (non-sequence-based systems do not explicitly calculate cycle, offset, and split).* *(Select requirements from both groups when the vendor is given the choice of supplying one type of adaptive operation or the other.)*

Requirements Document Reference Number	System Requirements Sample Requirements	Need Statement (Con Ops)
		4.1.0-1.0-4 Manage the lengths of queues *Note to user when selecting these requirements:* *Select from requirements in the 2.2 group when sequence-based systems are allowed (sequence-based systems explicitly calculate cycle, offset, and split).* *Select from requirements in the 2.3 group when non-sequence-based systems are allowed (non-sequence-based systems do not explicitly calculate cycle, offset, and split).* *(Select requirements from both groups when the vendor is given the choice of supplying one type of adaptive operation or the other.)*
2.2.0-5.0-4.0-1.0-2	(Sequence-based only) The increased limit shall be user-defined.	4.1.0-1.0-1 Maximize the throughput on coordinated routes *Note to user when selecting these requirements:* *Select from requirements in the 2.2 group when sequence-based systems are allowed (sequence-based systems explicitly calculate cycle, offset, and split).* *Select from requirements in the 2.3 group when non-sequence-based systems are allowed (non-sequence-based systems do not explicitly calculate cycle, offset, and split).* *(Select requirements from both groups when the vendor is given the choice of supplying one type of adaptive operation or the other.)*

Requirements Document Reference Number	System Requirements Sample Requirements	Need Statement (Con Ops)
		4.1.0-1.0-2 Provide smooth flow along coordinated routes *Note to user when selecting these requirements:* *Select from requirements in the 2.2 group when sequence-based systems are allowed (sequence-based systems explicitly calculate cycle, offset, and split).* *Select from requirements in the 2.3 group when non-sequence-based systems are allowed (non-sequence-based systems do not explicitly calculate cycle, offset, and split).* *(Select requirements from both groups when the vendor is given the choice of supplying one type of adaptive operation or the other.)* 4.1.0-1.0-3 Distribute phase times in an equitable fashion *Note to user when selecting these requirements:* *Select from requirements in the 2.2 group when sequence-based systems are allowed (sequence-based systems explicitly calculate cycle, offset, and split).* *Select from requirements in the 2.3 group when non-sequence-based systems are allowed (non-sequence-based systems do not explicitly calculate cycle, offset, and split).* *(Select requirements from both groups when the vendor is given the choice of supplying one type of adaptive operation or the other.)*

Requirements Document Reference Number	System Requirements Sample Requirements	Need Statement (Con Ops)
		4.1.0-1.0-4 Manage the lengths of queues *Note to user when selecting these requirements:* *Select from requirements in the 2.2 group when sequence-based systems are allowed (sequence-based systems explicitly calculate cycle, offset, and split).* *Select from requirements in the 2.3 group when non-sequence-based systems are allowed (non-sequence-based systems do not explicitly calculate cycle, offset, and split).* *(Select requirements from both groups when the vendor is given the choice of supplying one type of adaptive operation or the other.)*
2.2.0-5.0-5	(Sequence-based only) The ASCT shall adjust offsets to minimize the chance of stopping vehicles approaching a signal that have been served by a user-specified phase at an upstream signal.	**4.1.0-5** The system operator needs to minimize the chance that a queue forms at a specified location. *Note to user when selecting these requirements:* *Select from requirements in the 2.2 group when sequence-based systems are allowed (sequence-based systems explicitly calculate cycle, offset, and split).* *Select from requirements in the 2.3 group when non-sequence-based systems are allowed (non-sequence-based systems do not explicitly calculate cycle, offset, and split).* *Select from requirements in the 2.5 group when phase-based systems are allowed (phase-based systems do not explicitly calculate cycle, offset and split at all intersections).* *(Select requirements from two or all three groups when the vendor is given the choice of supplying the type of adaptive operation.)*
2.3	2.3 Non-sequence-based adaptive coordination	

Model Systems Engineering Documents for Adaptive Signal Control Technology (ASCT) Systems

Requirements Document Reference Number	System Requirements Sample Requirements	Need Statement (Con Ops)
2.3.0-1	Use this section if non-sequence-based adaptive coordination is likely to provide acceptable operation in your situation.	
2.3.0-2	(Non-sequence-based only) The ASCT shall calculate the appropriate state of the signal to suit the current coordination strategy at the critical signal controller. (A critical signal controller is defined by the user.)	4.1.0-1.0-1 Maximize the throughput on coordinated routes Note to user when selecting these requirements: Select from requirements in the 2.2 group when sequence-based systems are allowed (sequence-based systems explicitly calculate cycle, offset, and split). Select from requirements in the 2.3 group when non-sequence-based systems are allowed (non-sequence-based systems do not explicitly calculate cycle, offset, and split). (Select requirements from both groups when the vendor is given the choice of supplying one type of adaptive operation or the other.) 4.1.0-1.0-2 Provide smooth flow along coordinated routes Note to user when selecting these requirements: Select from requirements in the 2.2 group when sequence-based systems are allowed (sequence-based systems explicitly calculate cycle, offset, and split). Select from requirements in the 2.3 group when non-sequence-based systems are allowed (non-sequence-based systems do not explicitly calculate cycle, offset, and split). (Select requirements from both groups when the vendor is given the choice of supplying one type of adaptive operation or the other.)

Requirements Document Reference Number	System Requirements Sample Requirements	Need Statement (Con Ops)
		4.1.0-1.0-3 Distribute phase times in an equitable fashion Note to user when selecting these requirements: Select from requirements in the 2.2 group when sequence-based systems are allowed (sequence-based systems explicitly calculate cycle, offset, and split). Select from requirements in the 2.3 group when non-sequence-based systems are allowed (non-sequence-based systems do not explicitly calculate cycle, offset, and split). (Select requirements from both groups when the vendor is given the choice of supplying one type of adaptive operation or the other.) 4.1.0-1.0-4 Manage the lengths of queues Note to user when selecting these requirements: Select from requirements in the 2.2 group when sequence-based systems are allowed (sequence-based systems explicitly calculate cycle, offset, and split). Select from requirements in the 2.3 group when non-sequence-based systems are allowed (non-sequence-based systems do not explicitly calculate cycle, offset, and split). (Select requirements from both groups when the vendor is given the choice of supplying one type of adaptive operation or the other.)

Requirements Document Reference Number	System Requirements Sample Requirements	Need Statement (Con Ops)
2.3.0-3	(Non-sequence-based only) At non-critical intersections within a group, the ASCT shall calculate the time at which a user-specified phase shall be green, relative to a reference point at the critical intersection, to suit the current coordination strategy.	4.1.0-1.0-1 Maximize the throughput on coordinated routes *Note to user when selecting these requirements:* *Select from requirements in the 2.2 group when sequence-based systems are allowed (sequence-based systems explicitly calculate cycle, offset, and split).* *Select from requirements in the 2.3 group when non-sequence-based systems are allowed (non-sequence-based systems do not explicitly calculate cycle, offset, and split).* *(Select requirements from both groups when the vendor is given the choice of supplying one type of adaptive operation or the other.)* 4.1.0-1.0-2 Provide smooth flow along coordinated routes *Note to user when selecting these requirements:* *Select from requirements in the 2.2 group when sequence-based systems are allowed (sequence-based systems explicitly calculate cycle, offset, and split).* *Select from requirements in the 2.3 group when non-sequence-based systems are allowed (non-sequence-based systems do not explicitly calculate cycle, offset, and split).* *(Select requirements from both groups when the vendor is given the choice of supplying one type of adaptive operation or the other.)*

Requirements Document Reference Number	System Requirements Sample Requirements	Need Statement (Con Ops)
		4.1.0-1.0-3 Distribute phase times in an equitable fashion *Note to user when selecting these requirements:* *Select from requirements in the 2.2 group when sequence-based systems are allowed (sequence-based systems explicitly calculate cycle, offset, and split).* *Select from requirements in the 2.3 group when non-sequence-based systems are allowed (non-sequence-based systems do not explicitly calculate cycle, offset, and split).* *(Select requirements from both groups when the vendor is given the choice of supplying one type of adaptive operation or the other.)* 4.1.0-1.0-4 Manage the lengths of queues *Note to user when selecting these requirements:* *Select from requirements in the 2.2 group when sequence-based systems are allowed (sequence-based systems explicitly calculate cycle, offset, and split).* *Select from requirements in the 2.3 group when non-sequence-based systems are allowed (non-sequence-based systems do not explicitly calculate cycle, offset, and split).* *(Select requirements from both groups when the vendor is given the choice of supplying one type of adaptive operation or the other.)*

Requirements Document Reference Number	System Requirements Sample Requirements	Need Statement (Con Ops)
2.3.0-4	(Non-sequence-based only) When demand is present, the ASCT shall implement a user-specified maximum time between successive displays of each phase at each intersection.	4.1.0-1.0-1 Maximize the throughput on coordinated routes *Note to user when selecting these requirements:* *Select from requirements in the 2.2 group when sequence-based systems are allowed (sequence-based systems explicitly calculate cycle, offset, and split).* *Select from requirements in the 2.3 group when non-sequence-based systems are allowed (non-sequence-based systems do not explicitly calculate cycle, offset, and split).* *(Select requirements from both groups when the vendor is given the choice of supplying one type of adaptive operation or the other.)* 4.1.0-1.0-2 Provide smooth flow along coordinated routes *Note to user when selecting these requirements:* *Select from requirements in the 2.2 group when sequence-based systems are allowed (sequence-based systems explicitly calculate cycle, offset, and split).* *Select from requirements in the 2.3 group when non-sequence-based systems are allowed (non-sequence-based systems do not explicitly calculate cycle, offset, and split).* *(Select requirements from both groups when the vendor is given the choice of supplying one type of adaptive operation or the other.)*

Requirements Document Reference Number	System Requirements Sample Requirements	Need Statement (Con Ops)
		4.1.0-1.0-3 Distribute phase times in an equitable fashion Note to user when selecting these requirements: *Select from requirements in the 2.2 group when sequence-based systems are allowed (sequence-based systems explicitly calculate cycle, offset, and split).* *Select from requirements in the 2.3 group when non-sequence-based systems are allowed (non-sequence-based systems do not explicitly calculate cycle, offset, and split).* *(Select requirements from both groups when the vendor is given the choice of supplying one type of adaptive operation or the other.)* 4.1.0-1.0-4 Manage the lengths of queues Note to user when selecting these requirements: *Select from requirements in the 2.2 group when sequence-based systems are allowed (sequence-based systems explicitly calculate cycle, offset, and split).* *Select from requirements in the 2.3 group when non-sequence-based systems are allowed (non-sequence-based systems do not explicitly calculate cycle, offset, and split).* *(Select requirements from both groups when the vendor is given the choice of supplying one type of adaptive operation or the other.)*

Requirements Document Reference Number	System Requirements Sample Requirements	Need Statement (Con Ops)
2.3.0-5	(Non-sequence-based only) The ASCT shall adjust signal timing so that vehicles approaching a signal that have been served during a user-specified phase at an upstream signal do not stop.	4.1.0-5 The system operator needs to minimize the chance that a queue forms at a specified location. *Note to user when selecting these requirements:* *Select from requirements in the 2.2 group when sequence-based systems are allowed (sequence-based systems explicitly calculate cycle, offset, and split).* *Select from requirements in the 2.3 group when non-sequence-based systems are allowed (non-sequence-based systems do not explicitly calculate cycle, offset, and split).* *Select from requirements in the 2.5 group when phase-based systems are allowed (phase-based systems do not explicitly calculate cycle, offset and split at all intersections).* *(Select requirements from two or all three groups when the vendor is given the choice of supplying the type of adaptive operation.)*
2.4	**2.4 Single intersection adaptive operation**	
2.4.0-1	Use this section if non-coordinated adaptive coordination is likely to provide acceptable operation in your situation.	
2.4.0-2	The ASCT shall calculate a cycle length of a single intersection, based on current measured traffic conditions. (The calculation is based on the optimization objectives.)	4.1.0-1.0-6 At an isolated intersection, optimize operation with a minimum of phase failures (based on the optimization objectives).

Requirements Document Reference Number	System Requirements Sample Requirements	Need Statement (Con Ops)
2.4.0-3	The ASCT shall calculate optimum phase lengths, based on current measured traffic conditions. (The calculation is based on the optimization objectives.)	4.1.0-1.0-3 Distribute phase times in an equitable fashion *Note to user when selecting these requirements:* *Select from requirements in the 2.2 group when sequence-based systems are allowed (sequence-based systems explicitly calculate cycle, offset, and split).* *Select from requirements in the 2.3 group when non-sequence-based systems are allowed (non-sequence-based systems do not explicitly calculate cycle, offset, and split).* *(Select requirements from both groups when the vendor is given the choice of supplying one type of adaptive operation or the other.)* 4.1.0-1.0-6 At an isolated intersection, optimize operation with a minimum of phase failures (based on the optimization objectives).
2.4.03.0-1	The ASCT shall limit the difference between the length of a given phase and the length of the same phase during its next service to a user-specified value.	4.1.0-1.0-3 Distribute phase times in an equitable fashion *Note to user when selecting these requirements:* *Select from requirements in the 2.2 group when sequence-based systems are allowed (sequence-based systems explicitly calculate cycle, offset, and split).* *Select from requirements in the 2.3 group when non-sequence-based systems are allowed (non-sequence-based systems do not explicitly calculate cycle, offset, and split).* *(Select requirements from both groups when the vendor is given the choice of supplying one type of adaptive operation or the other.)* 4.1.0-1.0-6 At an isolated intersection, optimize operation with a minimum of phase failures (based on the optimization objectives).

Requirements Document Reference Number	System Requirements Sample Requirements	Need Statement (Con Ops)
2.4.03.0-2	When queues are detected at user-specified locations, the ASCT shall execute user-specified timing plan/operational mode.	4.1.0-1.0-3 Distribute phase times in an equitable fashion *Note to user when selecting these requirements:* *Select from requirements in the 2.2 group when sequence-based systems are allowed (sequence-based systems explicitly calculate cycle, offset, and split).* *Select from requirements in the 2.3 group when non-sequence-based systems are allowed (non-sequence-based systems do not explicitly calculate cycle, offset, and split).* *(Select requirements from both groups when the vendor is given the choice of supplying one type of adaptive operation or the other.)*
2.4.0.4	The ASCT shall calculate phase order, based on current measured traffic conditions. (The calculation is based on the optimization objectives.)	4.1.0-1.0-6 At an isolated intersection, optimize operation with a minimum of phase failures (based on the optimization objectives).
2.5	2.5 Phase-based adaptive coordination	4.1.0-1.0-6 At an isolated intersection, optimize operation with a minimum of phase failures (based on the optimization objectives).
2.5.0-1	Use this section if phase-based adaptive coordination is likely to provide acceptable operation in your situation.	
2.5.0-2	(Phase-based only) The ASCT shall alter the state of the signal controller for all phases at the user-specified intersection.	4.1.0-2 The system operator needs to manage the coordination in small groups of signals to link phase service at some intersections with phase service at adjacent intersections. *Note that phase-based systems do not explicitly calculate cycle, offset and split at all intersections.*

Requirements Document Reference Number	System Requirements Sample Requirements	Need Statement (Con Ops)
2.5.0-3	(Phase-based only) The ASCT shall calculate the time at which a user-specified phase shall be green at an intersection.	4.1.0-2 The system operator needs to manage the coordination in small groups of signals to link phase service at some intersections with phase service at adjacent intersections. *Note that phase-based systems do not explicitly calculate cycle, offset and split at all intersections.*
2.5.0-4	(Phase-based only) When demand is present, the ASCT shall implement a user-specified maximum time between successive displays of each phase at each intersection.	4.1.0-2 The system operator needs to manage the coordination in small groups of signals to link phase service at some intersections with phase service at adjacent intersections. *Note that phase-based systems do not explicitly calculate cycle, offset and split at all intersections.*
2.5.0-5	(Phase-based only) The ASCT shall alter the operation of the non-critical intersections to minimize stopping of traffic released from user-specified phases at the user-specified critical intersection.	4.1.0-2 The system operator needs to manage the coordination in small groups of signals to link phase service at some intersections with phase service at adjacent intersections. *Note that phase-based systems do not explicitly calculate cycle, offset and split at all intersections.*
2.5.0-6	(Phase-based only) The ASCT shall alter the operation of the non-critical intersections to minimize stopping of traffic arriving at user-specified phases at the user-specified critical intersection.	4.1.0-2 The system operator needs to manage the coordination in small groups of signals to link phase service at some intersections with phase service at adjacent intersections. *Note that phase-based systems do not explicitly calculate cycle, offset and split at all intersections.*

Requirements Document Reference Number	System Requirements Sample Requirements	Need Statement (Con Ops)
2.5.0.7	(Phase-based only) The ASCT shall adjust the state of the signal controller so that vehicles approaching a signal that have been served during a user-specified phase at an upstream signal do not stop.	4.1.0-5 The system operator needs to minimize the chance that a queue forms at a specified location. *Note to user when selecting these requirements:* *Select from requirements in the 2.2 group when sequence-based systems are allowed (sequence-based systems explicitly calculate cycle, offset, and split).* *Select from requirements in the 2.3 group when non-sequence-based systems are allowed (non-sequence-based systems do not explicitly calculate cycle, offset, and split).* *Select from requirements in the 2.5 group when phase-based systems are allowed (phase-based systems do not explicitly calculate cycle, offset and split at all intersections).* *(Select requirements from two or all three groups when the vendor is given the choice of supplying the type of adaptive operation.)* 4.1.0-2 The system operator needs to manage the coordination in small groups of signals to link phase service at some intersections with phase service at adjacent intersections. *Note that phase-based systems do not explicitly calculate cycle, offset and split at all intersections.*
2.6	2.6 Responsiveness	
2.6.0-1	The ASCT shall limit the change in consecutive cycle lengths to be less than a user-specified value.	4.8.0-1 The system operator needs to modify the ASCT operation to closely follow changes in traffic conditions.
2.6.0-2	The ASCT shall limit the change in phase times between consecutive cycles to be less than a user-specified value. (This does not apply to early gap-out or actuated phase skipping.)	4.8.0-1 The system operator needs to modify the ASCT operation to closely follow changes in traffic conditions.

Requirements Document Reference Number	System Requirements Sample Requirements	Need Statement (Con Ops)
2.6.0-3	The ASCT shall limit the changes in the direction of primary coordination to a user-specified frequency.	4.8.0-1 The system operator needs to modify the ASCT operation to closely follow changes in traffic conditions. 4.8.0-2 The system operator needs to constrain the selection of cycle lengths to those that provide acceptable operations, such as when resonant progression solutions are desired.
2.6.0-4	When a large change in traffic demand is detected, the ASCT shall respond more quickly than normal operation, subject to user-specified limits. (DEFINE "MORE QUICKLY")	4.8.0-3 The system operator needs to respond quickly to sudden large shifts in traffic conditions.
2.6.0-5	The ASCT shall select cycle length from a list of user-defined cycle lengths.	4.8.0-2 The system operator needs to constrain the selection of cycle lengths to those that provide acceptable operations, such as when resonant progression solutions are desired.
3	**3 External/Internal Interfaces**	

Requirements Document Reference Number	System Requirements Sample Requirements	Need Statement (Con Ops)
3.0-1	The ASCT shall support external interfaces according to the referenced interface control documents and the following detailed requirements. (Insert appropriate requirements that suit your needs. Interface data flows should be documented in your ITS architecture. Interface requirements include: • Information layer protocol • Application layer protocol • Lower layer protocol • Data aggregation • Frequency of storage • Frequency of reporting • Duration of storage)	4.3.0-1 The system operator needs to adaptively control signals operated by (specify jurisdictions). 4.3.0-2 The system operator needs to send data to another system that would allow the other system to coordinate with the ASCT system. 4.3.0-4 The system operator needs to receive data from another system that will allow the ASCT system to coordinate its operation with the adjacent system. 4.11.0-5 The system operator needs to report performance data in real time to (specify external system). 4.17.0-2 The system operator needs to react to commands issued by (specify an external control or decision support system, such as an ICM system or another signal system).
3.0-1.0-1	The ASCT shall send operational data to XX external system. (Insert appropriate requirements that suit your needs.)	4.3.0-2 The system operator needs to send data to another system that would allow the other system to coordinate with the ASCT system. 4.11.0-5 The system operator needs to report performance data in real time to (specify external system).
3.0-1.0-2	The ASCT shall send control data to the XX external system. (Insert appropriate requirements that suit your needs.)	4.3.0-2 The system operator needs to send data to another system that would allow the other system to coordinate with the ASCT system.
3.0-1.0-3	The ASCT shall send monitoring data to the XX external system. (Insert appropriate requirements that suit your needs.)	4.11.0-1 The agency needs the (specify external decision support system) to be able to monitor the ASCT system automatically.

Requirements Document Reference Number	System Requirements Sample Requirements	Need Statement (Con Ops)
3.0.1.0.4	The ASCT shall send coordination data to the XX external system. (Insert appropriate requirements that suit your needs.)	4.3.0-2 The system operator needs to send data to another system that would allow the other system to coordinate with the ASCT system.
3.0.1.0.5	The ASCT shall send performance data to the XX external system. (Insert appropriate requirements that suit your needs.)	4.11.0-5 The system operator needs to report performance data in real time to (specify external system).
3.0.1.0.6	The ASCT shall receive commands from the XX external system.	4.17.0-2 The system operator needs to react to commands issued by (specify an external control or decision support system, such as an ICM system or another signal system).
3.0.1.0.7	The ASCT shall implement the following commands from the XX external system when commanded: (Edit as appropriate for your situation) • Specified cycle length • Specified direction of progression • Specified adaptive strategy	4.17.0-2 The system operator needs to react to commands issued by (specify an external control or decision support system, such as an ICM system or another signal system).
4	**4 Crossing Arterials and Boundaries**	
4.0-1	The ASCT shall conform its operation to an external system's operation.	4.3.0-4 The system operator needs to receive data from another system that will allow the ASCT system to coordinate its operation with the adjacent system. 4.3.0-6 The system operator needs to detect traffic approaching from a neighboring system and coordinate the ASCT operation with the adjacent system. 4.17.0-2 The system operator needs to react to commands issued by (specify an external control or decision support system, such as an ICM system or another signal system).

Requirements Document Reference Number	System Requirements Sample Requirements	Need Statement (Con Ops)
4.0-1.0-1	The ASCT shall alter its operation to minimize interruption of traffic entering the system. (This may be achieved via detection, with no direct connection to the other system.)	4.3.0-4 The system operator needs to receive data from another system that will allow the ASCT system to coordinate its operation with the adjacent system. 4.3.0-6 The system operator needs to detect traffic approaching from a neighboring system and coordinate the ASCT operation with the adjacent system.
4.0-1.0-2	The ASCT shall operate a fixed cycle length to match the cycle length of an adjacent system.	4.3.0-5 The system operator needs to constrain the adaptive system to operate a cycle length compatible with the crossing arterial.
4.0-1.0-3	The ASCT shall alter its operation based on data received from another system.	4.3.0-4 The system operator needs to receive data from another system that will allow the ASCT system to coordinate its operation with the adjacent system.
4.0-1.0-4	The ASCT shall support adaptive coordination on crossing routes.	4.3.0-3 The system operator needs to adaptively coordinate signals on two crossing routes simultaneously. (Include signals on crossing arterials within the boundaries of the adaptive systems mapped in Chapter 3.)
5	**5 Access and Security**	
5.0-1	The ASCT shall be implemented with a security policy that addresses the following selected elements:	4.4.0-1 The system operator needs to have a security management and administrative system that allows access and operational privileges to be assigned, monitored and controlled by an administrator, and conform to the agency's access and network infrastructure security policies.
5.0.1.0-1	• Local access to the ASCT.	4.4.0-1 The system operator needs to have a security management and administrative system that allows access and operational privileges to be assigned, monitored and controlled by an administrator, and conform to the agency's access and network infrastructure security policies.

Requirements Document Reference Number	System Requirements Sample Requirements	Need Statement (Con Ops)
5.0-1.0-2	• Remote access to the ASCT.	4.4.0-1 The system operator needs to have a security management and administrative system that allows access and operational privileges to be assigned, monitored and controlled by an administrator, and conform to the agency's access and network infrastructure security policies.
5.0-1.0-3	• System monitoring.	4.4.0-1 The system operator needs to have a security management and administrative system that allows access and operational privileges to be assigned, monitored and controlled by an administrator, and conform to the agency's access and network infrastructure security policies.
5.0-1.0-4	• System manual override.	4.4.0-1 The system operator needs to have a security management and administrative system that allows access and operational privileges to be assigned, monitored and controlled by an administrator, and conform to the agency's access and network infrastructure security policies.
5.0-1.0-5	• Development	4.4.0-1 The system operator needs to have a security management and administrative system that allows access and operational privileges to be assigned, monitored and controlled by an administrator, and conform to the agency's access and network infrastructure security policies.
5.0-1.0-6	• Operations	4.4.0-1 The system operator needs to have a security management and administrative system that allows access and operational privileges to be assigned, monitored and controlled by an administrator, and conform to the agency's access and network infrastructure security policies.
5.0-1.0-7	• User login	4.4.0-1 The system operator needs to have a security management and administrative system that allows access and operational privileges to be assigned, monitored and controlled by an administrator, and conform to the agency's access and network infrastructure security policies.

Requirements Document Reference Number	System Requirements Sample Requirements	Need Statement (Con Ops)
5.0.1.0-8	• User password	4.4.0-1 The system operator needs to have a security management and administrative system that allows access and operational privileges to be assigned, monitored and controlled by an administrator, and conform to the agency's access and network infrastructure security policies.
5.0.1.0-9	• Administration of the system	4.4.0-1 The system operator needs to have a security management and administrative system that allows access and operational privileges to be assigned, monitored and controlled by an administrator, and conform to the agency's access and network infrastructure security policies.
5.0.1.0-10	• Signal controller group access	4.4.0-1 The system operator needs to have a security management and administrative system that allows access and operational privileges to be assigned, monitored and controlled by an administrator, and conform to the agency's access and network infrastructure security policies.
5.0.1.0-11	• Access to classes of equipment	4.4.0-1 The system operator needs to have a security management and administrative system that allows access and operational privileges to be assigned, monitored and controlled by an administrator, and conform to the agency's access and network infrastructure security policies.
5.0.1.0-12	• Access to equipment by jurisdiction	4.4.0-1 The system operator needs to have a security management and administrative system that allows access and operational privileges to be assigned, monitored and controlled by an administrator, and conform to the agency's access and network infrastructure security policies.
5.0.1.0-13	• Output activation	4.4.0-1 The system operator needs to have a security management and administrative system that allows access and operational privileges to be assigned, monitored and controlled by an administrator, and conform to the agency's access and network infrastructure security policies.

Requirements Document Reference Number	System Requirements Sample Requirements	Need Statement (Con Ops)
5.0-1.0-14	• System parameters	4.4.0-1 The system operator needs to have a security management and administrative system that allows access and operational privileges to be assigned, monitored and controlled by an administrator, and conform to the agency's access and network infrastructure security policies.
5.0-1.0-15	• Report generation	4.4.0-1 The system operator needs to have a security management and administrative system that allows access and operational privileges to be assigned, monitored and controlled by an administrator, and conform to the agency's access and network infrastructure security policies.
5.0-1.0-16	• Configuration	4.4.0-1 The system operator needs to have a security management and administrative system that allows access and operational privileges to be assigned, monitored and controlled by an administrator, and conform to the agency's access and network infrastructure security policies.
5.0-1.0-17	• Security alerts	4.4.0-1 The system operator needs to have a security management and administrative system that allows access and operational privileges to be assigned, monitored and controlled by an administrator, and conform to the agency's access and network infrastructure security policies.
5.0-1.0-18	• Security logging	4.4.0-1 The system operator needs to have a security management and administrative system that allows access and operational privileges to be assigned, monitored and controlled by an administrator, and conform to the agency's access and network infrastructure security policies.
5.0-1.0-19	• Security reporting	4.4.0-1 The system operator needs to have a security management and administrative system that allows access and operational privileges to be assigned, monitored and controlled by an administrator, and conform to the agency's access and network infrastructure security policies.

Requirements Document Reference Number	System Requirements Sample Requirements	Need Statement (Con Ops)
5.0-1.0-20	• Database	4.4.0-1 The system operator needs to have a security management and administrative system that allows access and operational privileges to be assigned, monitored and controlled by an administrator, and conform to the agency's access and network infrastructure security policies.
5.0-1.0-21	• Signal controller	4.4.0-1 The system operator needs to have a security management and administrative system that allows access and operational privileges to be assigned, monitored and controlled by an administrator, and conform to the agency's access and network infrastructure security policies.
5.0-2	The ASCT shall provide monitoring and control access at the following locations:	4.10.0-1 The system operator needs to monitor and control all required features of adaptive operation from the following locations: (Edit and select as appropriate to suit your situation.)
5.0-2.0-1	• Agency TMC	4.10.0-1.0-1 • Agency TMC
5.0-2.0-2	• Maintenance facility	4.10.0-1.0-2 • Maintenance facility
5.0-2.0-3	• Agency LAN or WAN	4.10.0-1.0-3 • Workstations on agency LAN or WAN located at (specify)
5.0-2.0-4	• Other agency TMC	4.10.0-1.0-4 • Other agency's TMC (specify)
5.0-2.0-5	• Local controller cabinets	4.10.0-1.0-5 • Local controller cabinets
5.0-2.0-6	• Maintenance vehicles	4.10.0-1.0-6 • Maintenance vehicles
5.0-2.0-7	• Remote locations via internet	4.10.0-1.0-7 • Remote locations (specify)

Requirements Document Reference Number	System Requirements Sample Requirements	Need Statement (Con Ops)
5.0-3	The ASCT shall comply with the agency's security policy as described in (specify appropriate policy document).	4.4.0-1 The system operator needs to have a security management and administrative system that allows access and operational privileges to be assigned, monitored and controlled by an administrator, and conform to the agency's access and network infrastructure security policies.
5.0-4	The ASCT shall not prevent access to the local signal controller database, monitoring or reporting functions by any installed signal management system.	4.10.0-2 The operator needs to access to the database management, monitoring and reporting features and functions of the signal controllers and any related signal management system from the access points defined for those system components.
6	**6 Data Log**	
6.0-1	The ASCT shall log the following events: (edit as appropriate)	4.11.0-6 The system operator needs to be able to report the exact state of signal timing and input data for a specified period, to allow historical analysis of the system operation.
6.0-1.0-1	Time-stamped vehicle phase calls	4.11.0-6 The system operator needs to be able to report the exact state of signal timing and input data for a specified period, to allow historical analysis of the system operation.
6.0-1.0-2	Time-stamped pedestrian phase calls	4.11.0-6 The system operator needs to be able to report the exact state of signal timing and input data for a specified period, to allow historical analysis of the system operation.
6.0-1.0-3	Time-stamped emergency vehicle preemption calls	4.11.0-6 The system operator needs to be able to report the exact state of signal timing and input data for a specified period, to allow historical analysis of the system operation.
6.0-1.0-4	Time-stamped transit priority calls	4.11.0-6 The system operator needs to be able to report the exact state of signal timing and input data for a specified period, to allow historical analysis of the system operation.

Requirements Document Reference Number	System Requirements Sample Requirements	Need Statement (Con Ops)
6.0-1.0-5	Time-stamped railroad preemption calls	4.11.0-6 The system operator needs to be able to report the exact state of signal timing and input data for a specified period, to allow historical analysis of the system operation.
6.0-1.0-6	Time-stamped start and end of each phase	4.11.0-6 The system operator needs to be able to report the exact state of signal timing and input data for a specified period, to allow historical analysis of the system operation.
6.0-1.0-7	Time-stamped controller interval changes	4.11.0-6 The system operator needs to be able to report the exact state of signal timing and input data for a specified period, to allow historical analysis of the system operation.
6.0-1.0-8	Time-stamped start and end of each transition to a new timing plan	4.11.0-6 The system operator needs to be able to report the exact state of signal timing and input data for a specified period, to allow historical analysis of the system operation.
6.0-2	The ASCT shall export its systems log in the following formats: (edit as appropriate) • MS Excel • Text • CVS • Open source SQL database	4.11.0-4 The system operator needs to store all operational data and signal timing parameters calculated by the adaptive system, and export selected data to (specify appropriate external system).
6.0-3	The ASCT shall store the event log for a minimum of XX days	4.11.0-4 The system operator needs to store all operational data and signal timing parameters calculated by the adaptive system, and export selected data to (specify appropriate external system).

Requirements Document Reference Number	System Requirements Sample Requirements	Need Statement (Con Ops)
6.0-4	The ASCT shall store results of all signal timing parameter calculations for a minimum of XX days.	4.11.0-2 The system operator needs to store and report data used to calculate signal timing and have the data available for subsequent analysis. 4.11.0-3 The system operator needs to store and report data that can be used to measure traffic performance under adaptive control.
6.0-5	The ASCT shall store the following measured data in the form used as input to the adaptive algorithm for a minimum of XX days: (edit as appropriate) • Volume • Occupancy • Queue length • Phase utilization • Arrivals in green • Green band efficiency	4.11.0-7 Have the ability to generate historic and real-time reports that effectively support operation, maintenance and reporting of system performance and traffic conditions. 4.11.0-2 The system operator needs to store and report data used to calculate signal timing and have the data available for subsequent analysis. 4.11.0-3 The system operator needs to store and report data that can be used to measure traffic performance under adaptive control.
6.0-6	The ASCT system shall archive all data automatically after a user-specified period not less than XX days.	4.11.0-4 The system operator needs to store all operational data and signal timing parameters calculated by the adaptive system, and export selected data to (specify appropriate external system).
6.0-7	The ASCT shall provide data storage for a system size of XX signal controllers. The data to be stored shall include the following: (edit as appropriate) • Controller state data • Reports • Log data • Security data • ASCT parameters • Detector status data	4.11.0-4 The system operator needs to store all operational data and signal timing parameters calculated by the adaptive system, and export selected data to (specify appropriate external system).

Requirements Document Reference Number	System Requirements Sample Requirements	Need Statement (Con Ops)
6.0-8	The ASCT shall calculate and report relative data quality including: • The extent data is affected by detector faults • Other applicable items	4.11.0-7 Have the ability to generate historic and real-time reports that effectively support operation, maintenance and reporting of system performance and traffic conditions.
6.0-9	The ASCT shall report comparisons of logged data when requested by the user: • Day-to-day, • Hour-to-hour • Hour of day to hour of day • Hour of week to hour of week • day of week to day week • Day of year to day of year	4.11.0-7 Have the ability to generate historic and real-time reports that effectively support operation, maintenance and reporting of system performance and traffic conditions.
6.0-10	The ASCT shall store data logs in a standard database (specify as appropriate).	4.11.0-4 The system operator needs to store all operational data and signal timing parameters calculated by the adaptive system, and export selected data to (specify appropriate external system).
6.0-11	The ASCT shall report stored data in a form suitable to provide explanations of system behavior to public and politicians and to troubleshoot the system.	4.11.0-7 Have the ability to generate historic and real-time reports that effectively support operation, maintenance and reporting of system performance and traffic conditions.
6.0-12	The ASCT shall store the following data in XX minute increments: (edit as appropriate) • Volume • Occupancy • Queue length	4.11.0-2 The system operator needs to store and report data used to calculate signal timing and have the data available for subsequent analysis.
7	**7 Advanced Controller Operation**	
7.0-1	When specified by the user, the ASCT shall serve a vehicle phase more than once for each time the coordinated phase is served.	4.9.0-1.0-1 • Service a phase more than once per cycle

Requirements Document Reference Number	System Requirements Sample Requirements	Need Statement (Con Ops)
7.0-2	The ASCT shall provide a minimum of XX phase overlaps.	4.9.0-1.0-2 • Operate at least XX overlap phases
7.0-3	The ASCT shall accommodate a minimum of XX phases at each signal	4.9.0-1.0-3 • Operate four rings, 16 phases and up to three phases per ring (Edit to suit your needs)
7.0-4	The ASCT shall accommodate a minimum of XX rings at each signal.	4.9.0-1.0-3 • Operate four rings, 16 phases and up to three phases per ring (Edit to suit your needs)
7.0-5	The ASCT shall accommodate a minimum of XX phases per ring	4.9.0-1.0-3 • Operate four rings, 16 phases and up to three phases per ring (Edit to suit your needs)
7.0-6	The ASCT shall provide a minimum of XX different user-defined phase sequences for each signal.	4.1.0-6 The system operator needs to modify the sequence of phases to support the various operational strategies. 4.9.0-1.0-4 • Permit different phase sequences under different traffic conditions
7.0-6.0-1	Each permissible phase sequence shall be user-assignable to any signal timing plan.	4.1.0-6 The system operator needs to modify the sequence of phases to support the various operational strategies. 4.9.0-1.0-4 • Permit different phase sequences under different traffic conditions
7.0-6.0-2	Each permissible phase sequence shall be executable by a time of day schedule.	4.1.0-6 The system operator needs to modify the sequence of phases to support the various operational strategies. 4.9.0-1.0-4 • Permit different phase sequences under different traffic conditions

Requirements Document Reference Number	System Requirements Sample Requirements	Need Statement (Con Ops)
7.06.0.3	Each permissible phase sequence shall be executable based on measured traffic conditions	4.1.0-6 The system operator needs to modify the sequence of phases to support the various operational strategies. 4.9.0-1.0.4 • Permit different phase sequences under different traffic conditions
7.0-7	The ASCT shall not prevent a phase/overlap output by time-of-day.	4.1.0-6 The system operator needs to modify the sequence of phases to support the various operational strategies.
7.0-8	The ASCT shall not prevent a phase/overlap output based on an external input.	4.1.0-6 The system operator needs to modify the sequence of phases to support the various operational strategies. 4.17.0-2 The system operator needs to react to commands issued by (specify an external control or decision support system, such as an ICM system or another signal system).
7.0-9	The ASCT shall not prevent the following phases to be designated as coordinated phases. (User to list all required phases.)	4.1.0-6 The system operator needs to modify the sequence of phases to support the various operational strategies. 4.9.0-1.0.9 • Allow any phase to be designated as the coordinated phase
7.0-10	The ASCT shall have the option for a coordinated phase to be released early based on a user-definable point in the phase or cycle. (User select phase or cycle.)	4.9.0-1.0-12 • Allow the coordinated phase to terminate early under prescribed traffic conditions
7.0-11	The ASCT shall not prevent the controller from displaying flashing yellow arrow left turn or right turn. (SELECT AS APPLICABLE)	4.9.0-1.0-15 • Use flashing yellow arrow to control permissive left turns and right turns.

Requirements Document Reference Number	System Requirements Sample Requirements	Need Statement (Con Ops)
7.0-12	The ASCT shall not prevent the local signal controller from performing actuated phase control using XX extension/passage timers as assigned to user-specified vehicle detector input channels in the local controller.	4.9.0-1.0-11 • Allow the controller to respond independently to individual lanes of an approach. This may be implemented in the signal controller using XX extension/passage timers, which may be assignable to each vehicle detector input channel. This may allow the adaptive operation to be based on data from a specific detector, or by excluding specific detectors.
7.0-12.0-1	The ASCT shall operate adaptively using user-specified detector channels.	4.9.0-1.0-11 • Allow the controller to respond independently to individual lanes of an approach. This may be implemented in the signal controller using XX extension/passage timers, which may be assignable to each vehicle detector input channel. This may allow the adaptive operation to be based on data from a specific detector, or by excluding specific detectors.
7.0-13	When adaptive operation is used in conjunction with normal coordination, the ASCT shall not prevent a controller serving a cycle length different from the cycles used at adjacent intersections.	4.9.0-1.0-16 • Service side streets and pedestrian phases at minor locations more often than at adjacent signals when this can be done without compromising the quality of the coordination. (E.g., double-cycle mid-block pedestrian crossing signals.)
7.0-14	(Describe requirements to suit other custom controller features that must be accommodated.)	4.9.0-1.0-8 • Accommodate the following custom features used by this agency (describe the features)
7.0-15	The ASCT shall operate adaptively with the following detector logic. (DESCRIBE THE CUSTOM LOGIC)	4.9.0-1.0-7 • Allow detector logic at an intersection to be varied depending on local signal states
8	**8 Pedestrians**	
8.0-1	When a pedestrian phase is called, the ASCT shall execute pedestrian phases up to XX seconds before the vehicle green of the related vehicle phase.	4.6.0-5 The system operator needs to accommodate early start of walk and exclusive pedestrian phases.
8.0-2	When a pedestrian phase is called, the ASCT shall accommodate pedestrian crossing times during adaptive operations.	4.6.0-2 The system operator needs to accommodate infrequent pedestrian operation while maintaining adaptive operation. (This is appropriate for pedestrian calls that are common but not so frequent that they drive the operational needs.)

Requirements Document Reference Number	System Requirements Sample Requirements	Need Statement (Con Ops)
8.0-3	When a pedestrian phase is called, the ASCT shall accommodate pedestrian crossing times then resume adaptive operation.	4.6.0-3 The system operator needs to incorporate frequent pedestrian operation into routine adaptive operation. (This is appropriate when pedestrians are frequent enough that they must be assumed to be present every cycle or nearly every cycle.)
8.0-4	The ASCT shall execute user-specified exclusive pedestrian phases during adaptive operation.	4.6.0-1 The system operator needs to accommodate infrequent pedestrian operation and then adaptively recover. (This is appropriate for rare pedestrian calls.)
8.0-5	The ASCT shall execute pedestrian recall on user-defined phases in accordance with a time of day schedule.	4.6.0-5 The system operator needs to accommodate early start of walk and exclusive pedestrian phases.
8.0-6	The ASCT shall begin a non-coordinated phase later than its normal starting point within the cycle when all of the following conditions exist: • The user enables this feature • Sufficient time in the cycle remains to serve the minimum green times for the phase and the subsequent non-coordinated phases before the beginning of the coordinated phase • The phase is called after its normal start time • The associated pedestrian phase is not called	4.6.0-3 The system operator needs to incorporate frequent pedestrian operation into routine adaptive operation. (This is appropriate when pedestrians are frequent enough that they must be assumed to be present every cycle or nearly every cycle.) 4.9.0-1.0-13 • Allow flexible timing of non-coordinated phases (such as late start of a phase) while maintaining coordination
8.0-7	When specified by the user, the ASCT shall execute pedestrian recall on pedestrian phase adjacent to coordinated phases.	4.6.0-3 The system operator needs to incorporate frequent pedestrian operation into routine adaptive operation. (This is appropriate when pedestrians are frequent enough that they must be assumed to be present every cycle or nearly every cycle.)

Requirements Document Reference Number	System Requirements Sample Requirements	Need Statement (Con Ops)
8.0-8	When the pedestrian phases are on recall, the ASCT shall accommodate pedestrian timing during adaptive operation.	4.6.0-3 The system operator needs to incorporate frequent pedestrian operation into routine adaptive operation. (This is appropriate when pedestrians are frequent enough that they must be assumed to be present every cycle or nearly every cycle.)
8.0-9	The ASCT shall not inhibit negative vehicle and pedestrian phase timing.	4.9.0-1.0-17 • Use negative pedestrian phasing to prevent an overlap conflicting with a pedestrian walk/don't walk
9	**9 Special Functions**	
9.0-1	The ASCT shall set a specific state for each special function output based on the occupancy on a user-specified detector.	4.9.0-1.0-11 • Allow the controller to respond independently to individual lanes of an approach. This may be implemented in the signal controller using XX extension/passage timers, which may be assignable to each vehicle detector input channel. This may allow the adaptive operation to be based on data from a specific detector, or by excluding specific detectors. 4.17.0-1 The system operator needs to be able to turn on signs that control traffic or provide driver information when specific traffic conditions occur, when needed to support the adaptive operation, when congestion is detected at critical locations or according to a time-of-day schedule
9.0-2	The ASCT shall set a specific state for each special function output based on the current cycle length.	4.17.0-1 The system operator needs to be able to turn on signs that control traffic or provide driver information when specific traffic conditions occur, when needed to support the adaptive operation, when congestion is detected at critical locations or according to a time-of-day schedule
9.0-3	The ASCT shall set a specific state for each special function output based on a time-of-day schedule.	4.17.0-1 The system operator needs to be able to turn on signs that control traffic or provide driver information when specific traffic conditions occur, when needed to support the adaptive operation, when congestion is detected at critical locations or according to a time-of-day schedule

Requirements Document Reference Number	System Requirements Sample Requirements	Need Statement (Con Ops)
10	**10 Detection**	
10.0-1	The ASCT shall be compatible with the following detector technologies (agency to specify): • Detector type A • Detector type B • Detector type C	
11	**11 Railroad and EV Preemption**	
11.0-1	The ASCT shall maintain adaptive operation at non-preempted intersections during railroad preemption.	4.13.0-1 The system operator needs to accommodate railroad and light rail preemption (explain further)
11.0-2	The ASCT shall maintain adaptive operation at non-preempted intersections during emergency vehicle preemption.	4.13.0-2 The system operator needs to accommodate emergency vehicle preemption (explain further)
11.0-3	The ASCT shall maintain adaptive operation at non-preempted intersections during Light Rail Transit preemption.	4.13.0-1 The system operator needs to accommodate railroad and light rail preemption (explain further)
11.0-4	The ASCT shall resume adaptive control of signal controllers when preemptions are released.	4.13.0-1 The system operator needs to accommodate railroad and light rail preemption (explain further) 4.13.0-2 The system operator needs to accommodate emergency vehicle preemption (explain further)
11.0-5	The ASCT shall execute user-specified actions at non-preempted signal controllers during preemption. (E.g., inhibit a phase, activate a sign, display a message on a DMS)	4.13.0-1 The system operator needs to accommodate railroad and light rail preemption (explain further) 4.13.0-2 The system operator needs to accommodate emergency vehicle preemption (explain further)

Requirements Document Reference Number	System Requirements Sample Requirements	Need Statement (Con Ops)
11.0-6	The ASCT shall operate normally at non-preempted signal controllers when special functions are engaged by a preemption event. (Examples of such special functions are a phase omit, a phase maximum recall or a fire route.)	4.13.0-1 The system operator needs to accommodate railroad and light rail preemption (explain further) 4.13.0-2 The system operator needs to accommodate emergency vehicle preemption (explain further)
11.0-7	The ASCT shall release user-specified signal controllers to local control when one signal in a group is preempted.	4.13.0-1 The system operator needs to accommodate railroad and light rail preemption (explain further) 4.13.0-2 The system operator needs to accommodate emergency vehicle preemption (explain further)
11.0-8	The ASCT shall not prevent the local signal controller from operating in normally detected limited-service actuated mode during preemption.	4.13.0-1 The system operator needs to accommodate railroad and light rail preemption (explain further) 4.13.0-2 The system operator needs to accommodate emergency vehicle preemption (explain further)
12	**12 Transit Priority**	
12.0-1	The ASCT shall continue adaptive operations of a group when one of its signal controllers has a transit priority call.	4.13.0-3 The system operator needs to accommodate bus and light rail transit signal priority (explain further)
12.0-2	The ASCT shall advance the start of a user-specified green phase in response to a transit priority call.	4.13.0-3 The system operator needs to accommodate bus and light rail transit signal priority (explain further)
12.0-2.0-1	The advance of start of green phase shall be user-defined.	4.13.0-3 The system operator needs to accommodate bus and light rail transit signal priority (explain further)
12.0-2.0-2	Adaptive operations shall continue during the advance of the start of green phase.	4.13.0-3 The system operator needs to accommodate bus and light rail transit signal priority (explain further)

Requirements Document Reference Number	System Requirements Sample Requirements	Need Statement (Con Ops)
12.0-3	The ASCT shall delay the end of a green phase, in response to a priority call.	4.13.0-3 The system operator needs to accommodate bus and light rail transit signal priority (explain further)
12.0-3.0-1	The delay of end of green phase shall be user-defined.	4.13.0-3 The system operator needs to accommodate bus and light rail transit signal priority (explain further)
12.0-3.0-2	Adaptive operations shall continue during the delay of the end of green phase.	4.13.0-3 The system operator needs to accommodate bus and light rail transit signal priority (explain further)
12.0-4	The ASCT shall permit at least XX exclusive transit phases.	4.13.0-3 The system operator needs to accommodate bus and light rail transit signal priority (explain further)
12.0-4.0-1	Adaptive operations shall continue when there is an exclusive transit phase call.	4.13.0-3 The system operator needs to accommodate bus and light rail transit signal priority (explain further)
12.0-5	The ASCT shall control vehicle phases independently of the following:	4.13.0-3 The system operator needs to accommodate bus and light rail transit signal priority (explain further)
12.0-5.0-1	• LRT only phases	4.13.0-3 The system operator needs to accommodate bus and light rail transit signal priority (explain further)
12.0-5.0-2	• Bus only phases	4.13.0-3 The system operator needs to accommodate bus and light rail transit signal priority (explain further)
12.0-6	The ASCT shall interface with external bus transit priority system in the following fashion..... (explain the external system and refer to other interfaces as appropriate)	4.13.0-3 The system operator needs to accommodate bus and light rail transit signal priority (explain further)
12.0-7	The ASCT shall interface with external light rail transit priority system in the following fashion..... (explain the external system and refer to other interfaces as appropriate)	4.13.0-3 The system operator needs to accommodate bus and light rail transit signal priority (explain further)

Requirements Document Reference Number	System Requirements Sample Requirements	Need Statement (Con Ops)
12.0-8	The ASCT shall accept a transit priority call from: • a signal controller/transit vehicle detector • an external system	4.13.0-3 The system operator needs to accommodate bus and light rail transit signal priority (explain further)
13	**13 Failure Events and Fallback**	
13.1	13.1 Detector Failure	
13.1.0-1	The ASCT shall take user-specified action in the absence of valid detector data from XX vehicle detectors within a group. (SELECT THE APPROPRIATE ACTION.)	4.14.0-1 The system operator needs to fall back to TOD or isolated free operation, as specified by the operator, without causing disruption to traffic flow, in the event of equipment, communications and software failure.
13.1.0-1.0-1	The ASCT shall release control to central system control.	4.14.0-1 The system operator needs to fall back to TOD or isolated free operation, as specified by the operator, without causing disruption to traffic flow, in the event of equipment, communications and software failure.
13.1.0-1.0-2	The ASCT shall release control to local operations to operate under its own time-of-day schedule.	4.14.0-1 The system operator needs to fall back to TOD or isolated free operation, as specified by the operator, without causing disruption to traffic flow, in the event of equipment, communications and software failure.
13.1.0-2	The ASCT shall use the following alternate data sources for operations in the absence of the real-time data from a detector:	4.14.0-1 The system operator needs to fall back to TOD or isolated free operation, as specified by the operator, without causing disruption to traffic flow, in the event of equipment, communications and software failure.
13.1.0-2.0-1	• Data from a user-specified alternate detector	4.14.0-1 The system operator needs to fall back to TOD or isolated free operation, as specified by the operator, without causing disruption to traffic flow, in the event of equipment, communications and software failure.

Requirements Document Reference Number	System Requirements Sample Requirements	Need Statement (Con Ops)
13.1.0-2.0-2	• Stored historical data from the failed detector	4.14.0-1 The system operator needs to fall back to TOD or isolated free operation, as specified by the operator, without causing disruption to traffic flow, in the event of equipment, communications and software failure.
13.1.0-2.0-3	The ASCT shall switch to the alternate source in real time without operator intervention.	4.14.0-1 The system operator needs to fall back to TOD or isolated free operation, as specified by the operator, without causing disruption to traffic flow, in the event of equipment, communications and software failure.
13.1.0-3	In the event of a detector failure, the ASCT shall issue an alarm to user-specified recipients. (This requirement may be fulfilled by sending the alarm to a designated list of recipients by a designated means, or by using an external maintenance management system.	4.12.0-1 The system operator needs to immediately notify maintenance and operations staff of alarms and alerts. 4.12.0-2 The system operator needs to immediately and automatically pass alarms and alerts to the (specify external system).
13.1.0-4	In the event of a failure, the ASCT shall log details of the failure in a permanent log.	4.12.0-3 The system operator needs to maintain a complete log of alarms and failure events.
13.1.0-5	The permanent failure log shall be searchable, archivable and exportable.	4.12.0-3 The system operator needs to maintain a complete log of alarms and failure events.
13.2	13.2 Communications Failure	
13.2-1	The ASCT shall execute user-specified actions when communications to one or more signal controllers fails within a group. {SELECT THE APPROPRIATE ACTION}	4.14.0-1 The system operator needs to fall back to TOD or isolated free operation, as specified by the operator, without causing disruption to traffic flow, in the event of equipment, communications and software failure.
13.2-1.0-1	In the event of loss of communication to a user-specified signal controller, the ASCT shall release control of all signal controllers within a user-specified group to local control.	4.14.0-1 The system operator needs to fall back to TOD or isolated free operation, as specified by the operator, without causing disruption to traffic flow, in the event of equipment, communications and software failure.

Requirements Document Reference Number	System Requirements Sample Requirements	Need Statement (Con Ops)
13.2-1.0-2	The ASCT shall switch to the alternate operation in real time without operator intervention.	4.14.0-1 The system operator needs to fall back to TOD or isolated free operation, as specified by the operator, without causing disruption to traffic flow, in the event of equipment, communications and software failure.
13.2-2	In the event of communications failure, the ASCT shall issue an alarm to user-specified recipients. (This requirement may be fulfilled by sending the alarm to a designated list of recipients by a designated means, or by using an external maintenance management system.	4.12.0-1 The system operator needs to immediately notify maintenance and operations staff of alarms and alerts. 4.12.0-2 The system operator needs to immediately and automatically pass alarms and alerts to the (specify external system).
13.2-3	The ASCT shall issue an alarm within XX minutes of detection of a failure.	4.12.0-1 The system operator needs to immediately notify maintenance and operations staff of alarms and alerts. 4.12.0-2 The system operator needs to immediately and automatically pass alarms and alerts to the (specify external system).
13.2-4	In the event of a communications failure, the ASCT shall log details of the failure in a permanent log.	4.12.0-3 The system operator needs to maintain a complete log of alarms and failure events.
13.2-5	The permanent failure log shall be searchable, archivable and exportable.	4.12.0-3 The system operator needs to maintain a complete log of alarms and failure events.
13.3	13.3 Adaptive Processor Failure	
13.3-1	The ASCT shall execute user-specified actions when adaptive control fails:	4.14.0-1 The system operator needs to fall back to TOD or isolated free operation, as specified by the operator, without causing disruption to traffic flow, in the event of equipment, communications and software failure.

Requirements Document Reference Number	System Requirements Sample Requirements	Need Statement (Con Ops)
13.3-1.0-1	The ASCT shall release control to central system control.	4.14.0-1 The system operator needs to fall back to TOD or isolated free operation, as specified by the operator, without causing disruption to traffic flow, in the event of equipment, communications and software failure.
13.3-1.0-2	The ASCT shall release control to local operations to operate under its own time-of-day schedule.	4.14.0-1 The system operator needs to fall back to TOD or isolated free operation, as specified by the operator, without causing disruption to traffic flow, in the event of equipment, communications and software failure.
13.3-2	In the event of adaptive processor failure, the ASCT shall issue an alarm to user-specified recipients. (This requirement may be fulfilled by sending the alarm to a designated list of recipients by a designated means, or by using an external maintenance management system.	4.12.0-1 The system operator needs to immediately notify maintenance and operations staff of alarms and alerts. 4.12.0-2 The system operator needs to immediately and automatically pass alarms and alerts to the (specify external system).
13.3-3	The permanent failure log shall be searchable, archivable and exportable.	
13.3-4	During adaptive processor failure, the ASCT shall provide all local detector inputs to the local controller.	4.14.0-1 The system operator needs to fall back to TOD or isolated free operation, as specified by the operator, without causing disruption to traffic flow, in the event of equipment, communications and software failure.
14	**14 Software**	
14.0-1	The vendor's adaptive software shall be fully operational within the following platform: (edit as appropriate) • Windows-PC • Linux • Mac-OS • Unix	4.15.0-2 The system operator needs to use equipment and software acceptable under current agency IT policies and procedures.
14.0-2	The ASCT shall fully satisfy all requirements when connected with detectors from manufacturer XX (specify required detector types).	4.15.0-1.0-2 • Detector type (list acceptable equipment)

Requirements Document Reference Number	System Requirements Sample Requirements	Need Statement (Con Ops)
14.0-3	The ASCT shall fully satisfy all requirements when connected with XX controllers (specify controller types).	4.15.0-1.0-1 • Controller type (list acceptable equipment)
15	**15 Training**	
15.0-1	The vendor shall provide the following training. (Edit as appropriate.)	4.16.0-1 The agency needs all staff involved in operation and maintenance to receive appropriate training.
15.0-1.0-1	The vendor shall provide training on the operations of the adaptive system.	4.16.0-1 The agency needs all staff involved in operation and maintenance to receive appropriate training.
15.0-1.0-2	The vendor shall provide training on troubleshooting the system.	4.16.0-1 The agency needs all staff involved in operation and maintenance to receive appropriate training.
15.0-1.0-3	The vendor shall provide training on preventive maintenance and repair of equipment.	4.16.0-1 The agency needs all staff involved in operation and maintenance to receive appropriate training.
15.0-1.0-4	The vendor shall provide training on system configuration.	4.16.0-1 The agency needs all staff involved in operation and maintenance to receive appropriate training.
15.0-1.0-5	The vendor shall provide training on administration of the system.	4.16.0-1 The agency needs all staff involved in operation and maintenance to receive appropriate training.
15.0-1.0-6	The vendor shall provide training on system calibration.	4.16.0-1 The agency needs all staff involved in operation and maintenance to receive appropriate training.
15.0-1.0-7	The vendor's training delivery shall include: printed course materials and references, electronic copies of presentations and references.	4.16.0-1 The agency needs all staff involved in operation and maintenance to receive appropriate training.
15.0-1.0-8	The vendor's training shall be delivered at (specify locations for training).	4.16.0-1 The agency needs all staff involved in operation and maintenance to receive appropriate training.
15.0-1.0-9	The vendor shall provide a minimum of XX hours training to a minimum of XX staff. (specify how much training will be required)	4.16.0-1 The agency needs all staff involved in operation and maintenance to receive appropriate training.

Requirements Document Reference Number	System Requirements Sample Requirements	Need Statement (Con Ops)
15.0-1.0-10	The vendor shall provide a minimum of XX training sessions (specify how many sessions over what period).	4.16.0-1 The agency needs all staff involved in operation and maintenance to receive appropriate training.
16	**16 Maintenance, Support and Warranty**	
16.0-1	The Maintenance Vendor shall provide maintenance according to a separate maintenance contract. That contract should identify repairs necessary to preserve requirements fulfillment, responsiveness in effecting those repairs, and all requirements on the maintenance provider while performing the repairs.	4.16.0-2 The agency needs the system to fulfill all requirements for the life of the system. The agency therefore needs the system to be maintained to repair faults that are not defects in materials and workmanship.
16.0-2	The Vendor shall provide routine updates to the software and software environment necessary to preserve the fulfillment of requirements for a period of XX years. Preservation of requirements fulfillment especially includes all IT management requirements as previously identified.	4.16.0-4 The agency needs the system to fulfill all requirements for the life of the system. The agency therefore needs support to keep software and software environment updated as necessary to prevent requirements no longer being fulfilled.
16.0-3	The Vendor shall warrant the system to be free of defects in materials and workmanship for a period of XX years. Warranty is defined as correcting defects in materials and workmanship (subject to other language included in the purchase documents). Defect is defined as any circumstance in which the material does not perform according to its specification.	4.16.0-3 The agency needs the system to fulfill all requirements for the life of the system. The agency therefore needs the system to remain free of defects in materials and workmanship that result in requirements no longer being fulfilled.
17	**17 Schedule**	
17.0-1	The ASCT shall set the state of external input/output states according to a time-of-day schedule.	4.17.0-1 The system operator needs to be able to turn on signs that control traffic or provide driver information when specific traffic conditions occur, when needed to support the adaptive operation, when congestion is detected at critical locations or according to a time-of-day schedule
17.0-2	The ASCT output states shall be settable according to a time-of-day schedule	4.17.0-1 The system operator needs to be able to turn on signs that control traffic or provide driver information when specific traffic conditions occur, when needed to support the adaptive operation, when congestion is detected at critical locations or according to a time-of-day schedule

Requirements Document Reference Number	System Requirements Sample Requirements	Need Statement (Con Ops)
17.0-3	The ASCT operational parameters shall be settable according to a Time of Day schedule	
18	**18 Performance Measurement, Monitoring and Reporting**	
18.0-1	The ASCT shall report measures of current traffic conditions on which it bases signal state alterations.	4.11.0-2 The system operator needs to store and report data used to calculate signal timing and have the data available for subsequent analysis.
18.0-2	The ASCT shall report all intermediate calculated values that are affected by calibration parameters.	4.11.0-2 The system operator needs to store and report data used to calculate signal timing and have the data available for subsequent analysis.
18.0-3	The ASCT shall maintain a log of all signal state alterations directed by the ASCT.	4.11.0-7 Have the ability to generate historic and real-time reports that effectively support operation, maintenance and reporting of system performance and traffic conditions. 4.11.0-2 The system operator needs to store and report data used to calculate signal timing and have the data available for subsequent analysis.
18.0.3.0-1	The ASCT log shall include all events directed by the external inputs.	4.11.0-7 Have the ability to generate historic and real-time reports that effectively support operation, maintenance and reporting of system performance and traffic conditions.
18.0.3.0-2	The ASCT log shall include all external output state changes.	4.11.0-7 Have the ability to generate historic and real-time reports that effectively support operation, maintenance and reporting of system performance and traffic conditions.
18.0.3.0-3	The ASCT log shall include all actual parameter values that are subject to user-specified values.	4.11.0-7 Have the ability to generate historic and real-time reports that effectively support operation, maintenance and reporting of system performance and traffic conditions.

Requirements Document Reference Number	System Requirements Sample Requirements	Need Statement (Con Ops)
18.0.3.0.4	The ASCT shall maintain the records in this ASCT log for XX period.	4.11.0.7 Have the ability to generate historic and real-time reports that effectively support operation, maintenance and reporting of system performance and traffic conditions.
18.0.3.0.5	The ASCT shall archive the ASCT log in the following manner: (Specify format, frequency, etc., to suit your needs.)	4.11.0.7 Have the ability to generate historic and real-time reports that effectively support operation, maintenance and reporting of system performance and traffic conditions.

www.ingramcontent.com/pod-product-compliance
Lightning Source LLC
Chambersburg PA
CBHW080237180526
45167CB00006B/2319